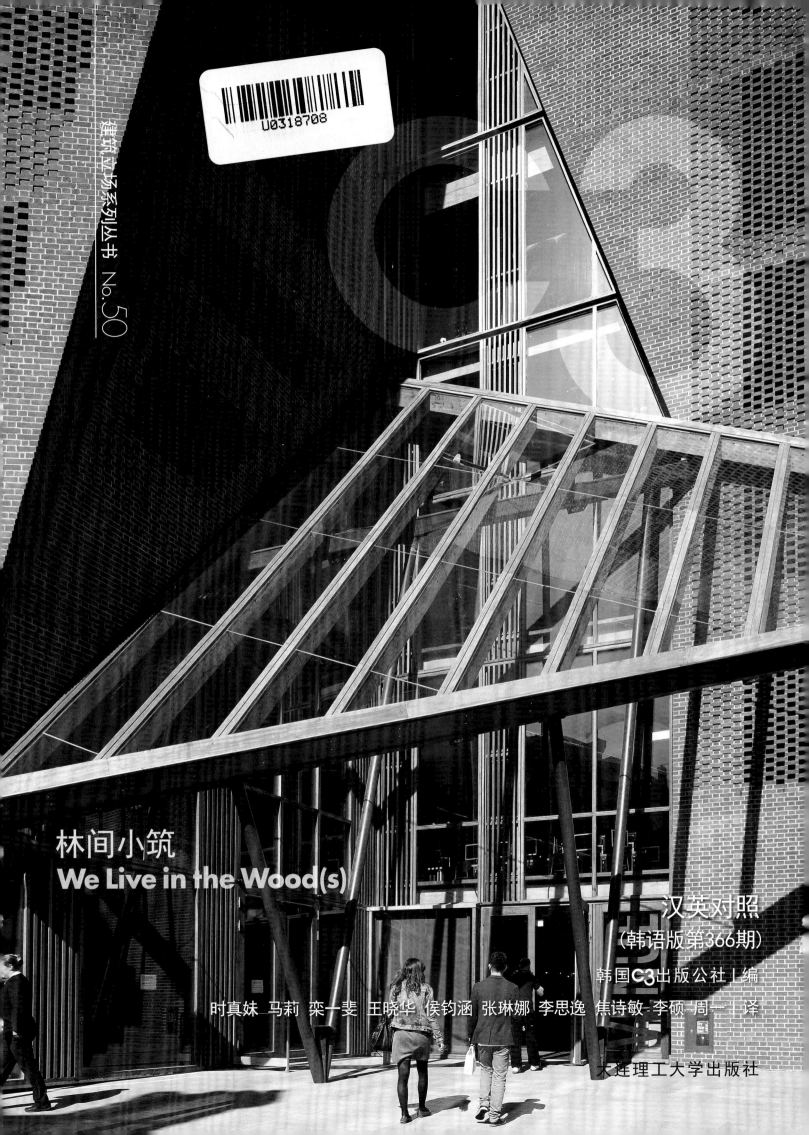

建筑立场系列丛书 No.50

林间小筑
We Live in the Wood(s)

汉英对照
（韩语版第366期）

韩国C3出版公社 | 编

时真妹 马莉 栾一斐 王晓华 侯钧涵 张琳娜 李思逸 焦诗敏 李硕 周一 | 译

大连理工大学出版社

| | |
|---|---|
| **4** | 004 新北西兰岛医院_Herzog & de Meuron |
| | 008 文叙瑟尔医院_C.F. Møller Architects |
| | 012 洛桑医科大学儿童医院_von Gerkan, Marg and Partner |

城市建造

## **16** 融入历史肌理

016 嵌入的建筑：融入历史肌理_Angelos Psilopoulos

020 百代基金会_Renzo Piano Building Workshop

032 伦敦政治经济学院学生中心_O'Donnell + Tuomey Architects

046 索纳·雷德大楼_Steven Holl Architects

住宅建造

## **60** 林间小筑

060 林间小筑_Diego Terna

066 Norderhov林中小屋_Atelier Oslo

076 梅里之家_Pezo Von Ellrichshausen Architects

088 "狗跑"住宅_Dunn & Hillam Architects

098 踢马公寓
  _Bohlin Cywinski Jackson in association with Bohlin Grauman Miller Architects

106 布鲁恩别墅_Häkli Architects

114 环绕红桉的住宅_MORQ

## **124** 卡萨格兰实验室

124 与自然共鸣

| | |
|---|---|
| 130 台东废墟学院 | 138 台北废墟学院 |
| 144 终极废墟 | 152 财富山 |
| 156 波将金公园 | 164 陈宅 |
| 170 昆虫屋 | 174 乌尼岛夏季剧院 |
| 176 景观 | 178 牡蛎人 |
| 182 Paracity项目 | |

188 建筑师索引

C3 建筑立场系列丛书 No. 50

**4**
004 New North Zealand Hospital _ Herzog & de Meuron
008 Vendsyssel Hospital _ C.F. Møller Architects
012 Lausanne University Hospital _ von Gerkan, Marg and Partner

Urban How
# Into the Historic Fabric
**16**
016 *Architectural Insertions: Building into the Historic Fabric _ Angelos Psilopoulos*
020 Pathé Foundation _ Renzo Piano Building Workshop
032 Student Center at the London School of Economics and Political Science
    _ O'Donnell + Tuomey Architects
046 Seona Reid Building _ Steven Holl Architects

Dwell How
# We Live in the Wood(s)
**60**
060 *We Live in the Wood(s) _ Diego Terna*
066 Cabin Norderhov _ Atelier Oslo
076 Meri House _ Pezo Von Ellrichshausen Architects
088 Dogtrot House _ Dunn & Hillam Architects
098 Kicking Horse Residence
    _ Bohlin Cywinski Jackson in association with Bohlin Grauman Miller Architects
106 Villa Bruun _ Häkli Architects
114 Karri Loop House _ MORQ

# Casagrande Laboratory
**124**
124 *Resonating with Nature*
130 Taitung Ruin Academy      138 Taipei Ruin Academy
144 Ultra-Ruin                152 Treasure Hill
156 Potemkin                  164 Chen House
170 Bug Dome                  174 Uunisaari Summer Theater
176 Land(e)scape              178 Oystermen
182 Paracity

188 Index

# 公共医疗机构 Public Hospitals

## 新北西兰岛医院 _Herzog & de Meuron

丹麦新北西兰岛医院最终选定赫尔佐格和德梅隆建筑事务所来负责新医院大楼的设计工作,项目位于首府哥本哈根的北部,毗邻希勒罗德镇。

建筑为自然所环绕,中央有一座巨大的花园。建筑的水平形态及波浪起伏的体量回应了位于丹麦中部大片景观中的项目场地。水平形态的建筑对于医院也是最合适的建筑类型,因为它有利于促进内部的沟通与交流:在此工作的人们,无论属于哪个科室,都有一个共同的目标:治愈病患。新医院将超越传统医院的运营界限。在过去的几十年里,那些楼体较高的医院很难实现这一目标。

设计方案将两个看似矛盾的目标糅合到了一起:在尽可能地拉近内部功能区之间联系的必要前提下,还满足了委托人对于建造一座大型中央花园的设想。折衷的方案是将建筑整体设计为一个十字形的有机体,让内部的花园随意穿流其间。花园下方的中央大厅被划割成了四个圆形的小型庭院。

由建筑剖面图可以看出,医院的功能布局被简化:两层的建筑体主要用来布置检查区和治疗区,成为基座,病房在此基座上呈两层带状结构沿周边分布,环绕形成了中央的巨大花园。基座之上,每个楼层都有多条不同线路与户外连通。庭院提供采光,还可作为观景点,让人在此轻松定位方向。大面积的连通区域、重复设置的内部天庭、统一规模的房间设计,确保了建筑的高度灵活性。

景观美化方案包含了两种典型的丹麦景观类型。周边由一个森林公园紧紧环抱着建筑,公园内还设有停车场,而内部的中央花园则

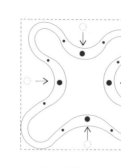

一座大型中央花园
one large central garden

连接室内的短距离垂直流线
vertical circulation
for short internal connections

病房在此基座上呈两层带状结构沿周边分布，环绕形成了中央的巨大花园
two floors of examination and treatment form a pedestal upon which a two storey ribbon of wards is placed along the perimeter forming the large central garden

种满了石楠灌木丛。环形的树丛形成了天然屏障，保护了诊疗区的隐私，同时也形成了花园小径网络。

该项目将成为希勒罗德南部区域新整体规划的中心。

### New North Zealand Hospital

The New North Zealand Hospital has appointed Herzog & de Meuron to design their new building, located north of Copenhagen, close to the town of Hillerød. The hospital is surrounded by nature and contains a garden in its center. The horizontality of the building and its undulating form respond to the location of the hospital in the midst of the wide Danish landscape. A horizontal building is an appropriate building typology for a hospital, because this fosters exchange: across the various departments, the employees work on a shared goal: the healing of the ailing human being. The new hospital shall over-

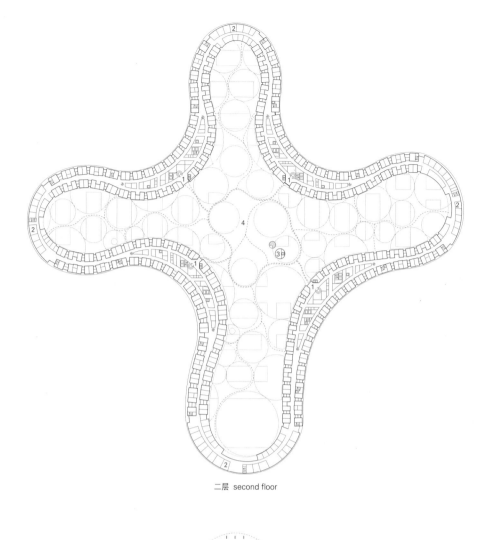

come conventional operational borders. The tall hospitals of the last decades have rarely achieved this goal.

The plan is the marriage of two seemingly contradictory goals: the desire for a large central garden and the necessity for short internal connections. The result is an organic cross shape that permits the interior garden to become a flowing space. A central hall underneath the garden is characterized by four round courtyards. Viewed in section, the arrangement of the functions is simple: two floors for examination and treatment form a pedestal upon which a two-story ribbon of wards is placed along the perimeter, forming the large central garden. In the pedestal, on both levels, connections to the outdoors are created in various ways. Courtyards provide daylight and vistas, and ease orientation. Large connected areas, the repetitive arrangement of the interior courtyards, and uniform room sizes offer a high degree of flexibility.

The landscaping concept consists of two typical Danish landscape typologies. A forest park with clearings for parking lots surrounds the building, while the central garden is Heathland. Circular hedges inhibit views into the pedestal and create the path network of the garden.

This project will be the heart of the new masterplan of South Hillerød.

二层 second floor
一层 first floor

1 双人病房   1. double ward
2 办公室   2. office
3 教堂   3. chapel
4 中央花园   4. central garden
5 接待处/大堂   5. reception/lobby
6 食堂   6. canteen
7 门诊诊所   7. outpatient clinics
8 治疗室   8. therapy
9 实验室   9. laboratories
10 透析室   10. dialysis
11 急诊部病房   11. emergency department ward
12 调查部   12. research
13 办公室   13. office
14 服务部   14. service village

# 文叙瑟尔医院 _C.F. Møller Architects

文叙瑟尔医院扩建和翻新项目的获奖设计,旨在寻求提高医院的整体建筑认同感,以及多方面整合建筑与景观,包括在屋顶建造的一处儿童游乐场。

这项委托任务包括建造一个新的治疗设施(占地面积约14 000m²),一处母婴护理中心,以及对文叙瑟尔医院场地(包括已有建筑)的整体扩建及翻新。

新翼楼的设计为封闭式,并将中部区域围成了一个采光良好的大型庭院。结构整体通透;一条环形流线连接着各个独立诊疗区,促进了各相邻功能区之间沟通的灵活性及重合性。实用且功能性强的连接使人们很容易在建筑周围找到出路。

轮廓清晰且与众不同的建筑采用了高标准的材料,且与带有一系列混合植被、石板路及雨水收集装置的室外空间景观的着重强调相结合,成功地利用了一切有利于改善患者治疗环境的因素。

建筑特色之一是三楼儿科病区的布局,为明亮的儿童友好型室内设计,一处封闭式的、有防护的游乐场,像一处绿洲,位于屋顶。

评审委员会的获奖理由陈述是:"该项设计的整体精神理念是所有参赛作品中最好的。其中的那些庭院,无论是内部的还是那些由建筑间衔接形成的,都是极其理想的。三楼隐蔽的室外空间能够提供极佳的视野及充足的自然光线,良好的环境对患者的治疗来说十分有益,甚至可以说建筑本身就成为了治愈过程中的一部分。

两侧较低楼层都选用了深色砖,带有空心玻璃立面的三楼/屋顶花园设计优雅,它们与一面由金属型材制成的围墙,共同促成了建筑比例及材料之间的有机互动,使医疗区与其他

建筑完美地协调一致。设计方案为新建筑以及原有建筑的未来发展奠定了同样令人折服的基调及个性。"

新建筑将实现2020年低能耗标准（根据丹麦建筑规范），达到每平方米少于25千瓦时能源使用的标准，并将得到丹麦绿色建筑委员会的银级认证。文叙瑟尔医院的新综合设施计划于2019年初竣工。

## Vendsyssel Hospital

The competition-winning design for the extension and refurbishment of the Vendsyssel Hospital seeks to enhance the overarching architectonic identity of the hospital and to integrate buildings and landscapes in various ways, including a rooftop children's oasis.

The task includes the construction of a new treatment facility (covering about 14,000m²), a mother and child unit and the overall planning of the extension and refurbishment of the Vendsyssel Hospital site including existing buildings.

The design of the new wings creates an enclosed form surrounding large, well-lit courtyards. The structure is transparent; a circulation loop connects the individual treatment areas and allows for flexibility and overlaps between neighbouring functions. There are efficient and highly functional links making it easy to find your way around the building.

The clear-cut yet distinctive architecture features sterling materials, in combination with a strong emphasis on landscape, including a series of outdoor spaces with integrated planting, paving and stormwa-

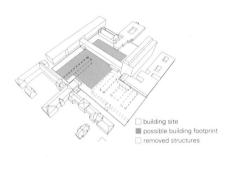

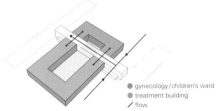

U形翼楼由横跨的大堂进行连接，以形成较短的路线和方便通行的通道
the U-shaped wings are connected across the lobby to create short routes and convenient accesses

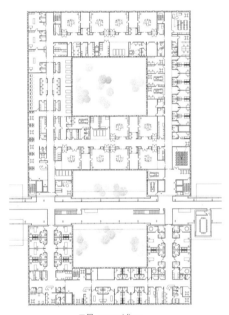

二层 second floor

三层 third floor

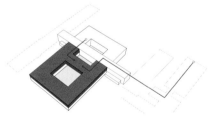

用作操作室的其他区域位于一层的十字连接区，确保应急情况下有效且简短的流线
the extra floor area required for operation rooms is provided in a cross connection on the 1st floor, which ensures short and efficient flows for emergencies

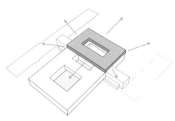

病房区位于顶层，保证了望向景观的视野和隐私。
the bed wards are placed on the top floor to ensure views to landscape and privacy

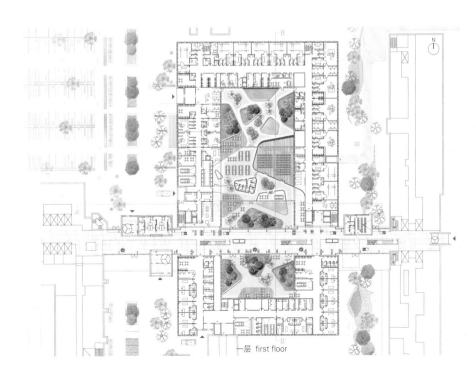

一层 first floor

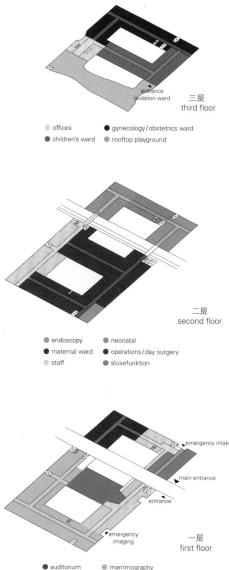

三层 third floor
- offices
- children's ward
- gynecology/obstetrics ward
- rooftop playground

二层 second floor
- endoscopy
- maternal ward
- staff
- neonatal
- operations/day surgery
- slusefunktion

一层 first floor
- auditorium
- imaging
- kiosk
- mammography
- Gyn/OBS ambulatory
- children's ambulatory

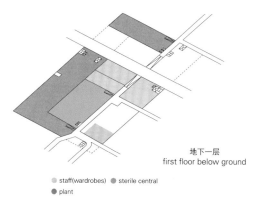

地下一层 first floor below ground
- staff(wardrobes)
- sterile central
- plant

ter handling, successfully exploiting every opportunity to create healing architecture. One special characteristic of the building is the layout of the third-floor paediatric unit, which includes a bright and children-friendly interior design, and an enclosed and protected playground, resembling a green oasis, on the roof.

The panel of judges stated: "The overall ethos of the design was best of all those received. The courtyards, both the internal ones and those that form connections between the buildings, are ideal. The secluded outdoor areas on the third floor, which have excellent views and plenty of natural sunlight, are convincing features which underpin the desire for the architecture to become part of the healing aspect of patients' treatment. The choice of dark bricks on both lower floors and the elegant design of the third floor/roof garden with airy glass facades and an espalier made of metal profiles create an elegant interplay of proportions and materials, coordinating perfectly with the Medical Unit and other buildings. The design proposal sets a convincing tone and character for the future development of new and existing buildings alike."

The new buildings will be low-energy class 2020 (by Danish Building Codes), which signifies a primary energy demand of less than 25 kWh/m², and are expected to be certified in accordance with Green Building Council Denmark's DGNB-DK Silver. The new complex is planned for completion in early 2019.

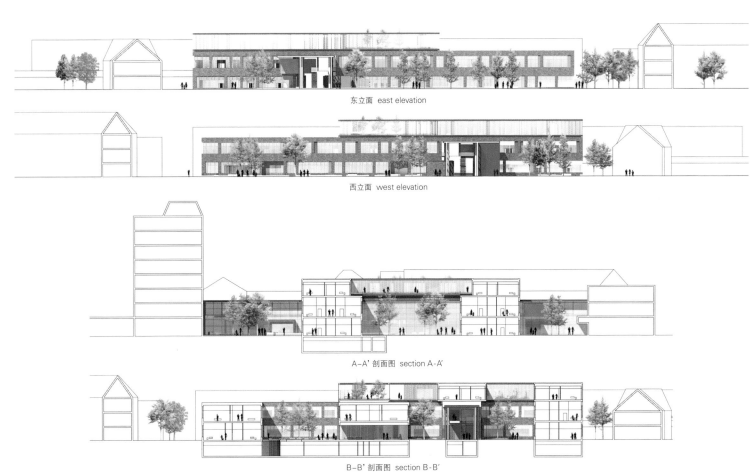

东立面 east elevation

西立面 west elevation

A-A' 剖面图 section A-A'

B-B' 剖面图 section B-B'

# 洛桑医科大学儿童医院 _von Gerkan, Marg and Partner

冯·格康,玛格及合伙人建筑事务所与JB Ferrari建筑事务所合作,为洛桑医科大学的一个新儿童医院设计方案,并最终赢得了国际大赛的一等奖。医院提供85个床位,并设有一个儿童急诊中心,暂定于2019年开放。设计理念包括可自然采光的绿化公共区域,尽可能使患者及家属在这里感到舒适愉快。目前,该项目的入口已经展现在洛桑的民众面前。

未来的儿童医院位于大学附属医院的中心位置,地处一处180m长的基地上,并处于Rue du Bugnon大街和广场交会的街角处,而这里也是城市规划中极其显要的位置。主体已经确定为六层楼的建筑,主入口设置在这处广场,可以通到儿童医院、急救中心以及护理站。此处原有的地铁站也将被整合到整体规划方案之中。广场东部一侧的建筑相对较低,这样就不会遮住位于其后的妇产医院的视野。

连接较低建筑和妇产医院的主通道充当了建筑内部的交通要道。一座宽敞明亮的二层建筑体,内部将容纳候诊和游乐区域,其间呈锯齿状分布着诊断区和治疗区。较低的建筑屋顶上设有咖啡厅,一个开放的绿化平台可提供游乐和休憩的空间。由于地形的优势,屋面平台高度较低,可从地面或经由一条天然的坡道到达;另外,沿广场内一侧外立面设置的宽大阶梯也可通到平台。

护理站位于六层高的建筑西翼区内,呈阶梯式的室内庭院和温室提升了空间的品质,同时也创造了受保护的室外活动空间。这里可以享受眺望洛桑城市、日内瓦湖以及阿尔卑斯山的壮阔视野。

## Lausanne University Hospital

The design by von Gerkan, Marg and Partner, with JB Ferrari, for a new children's hospital at the Lausanne University Hospital has won first prize in an international competition. The hospital is to provide 85 beds and an accident and emergency unit for children, and is provisionally scheduled to open in 2019. The design concept includes light-flooded open areas with much greenery, which are intended to make a hospital stay as agreeable as possible for patients and their relatives. Today, the entry will be presented to the public in Lausanne.

The future children's hospital will be located in the center of the University Hospital on a 180 meter long site on the corner of Rue du Bugnon and Esplanade which is a prominent location from the urban design point of view. The building has been designed with six stories, with the main entrance from Esplanade leading into the children's hospital, the accident

城市整合——建筑正面
urban integration – building fronts

无阻碍的视野——妇产/儿童门诊
unobstructed views – maternity/child clinic

绿色走廊
green corridor

公园
park

and emergency unit and the care wards. The underground station in that locality will be included in the overall concept by being built over. The part of the building to the east of the Esplanade will be lower, so that the view from the gynaecological unit would not be obstructed.

The main access avenue between this lower building and the gynaecological unit will serve as internal access. The generous, bright two-story building will accommodate waiting and play areas, with the examination and treatment areas branching off in a comb-like layout. On the roof of the lower building is the cafeteria, and a public planted terrace offers facilities for play and recreation. Thanks to the topography of the location, the terrace can be accessed both at ground level and via natural ramps; in addition, generous steps will lead from the Esplanade along the facade to the terrace.

The care wards in the six-story western part of the building will feature a terraced inner courtyard with conservatory which is intended as a protected outside play area. The location offers fantastic views of Lausanne, Lake Geneva and the Alps.

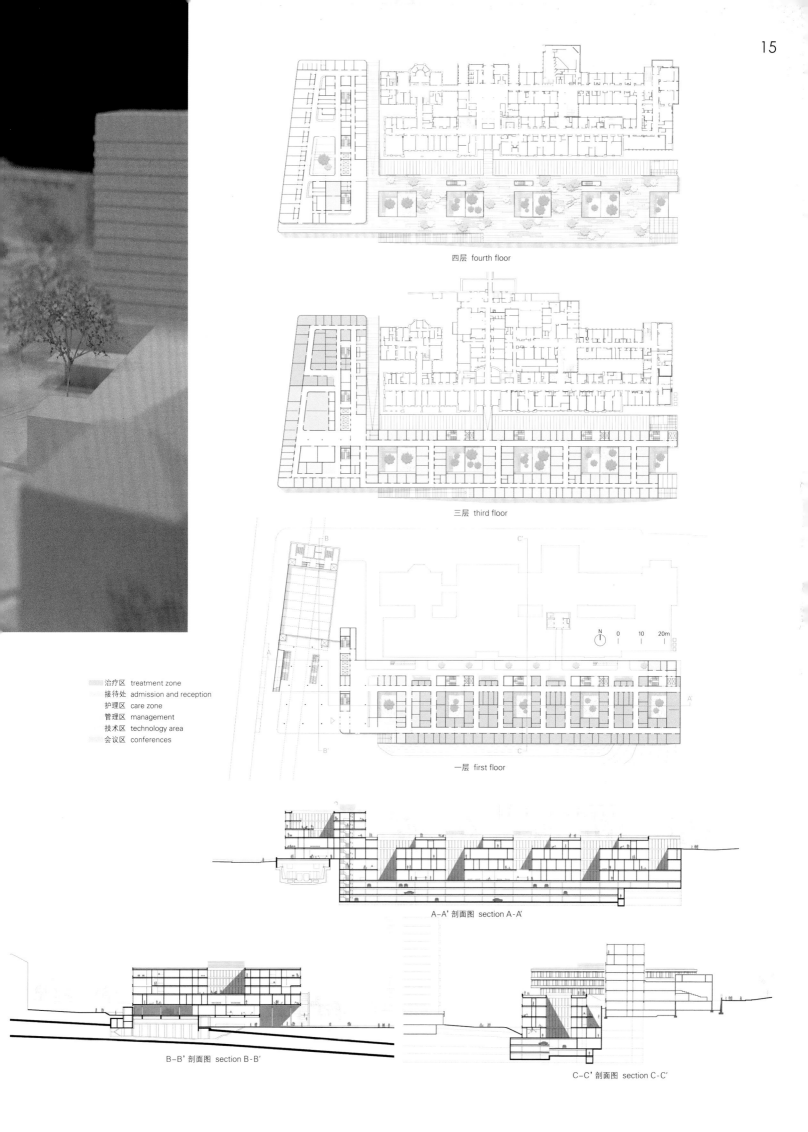

治疗区 treatment zone
接待处 admission and reception
护理区 care zone
管理区 management
技术区 technology area
会议区 conferences

四层 fourth floor
三层 third floor
一层 first floor

A-A' 剖面图 section A-A'
B-B' 剖面图 section B-B'
C-C' 剖面图 section C-C'

融入历史肌理

# Into the Historic

将建筑融入历史肌理在很大程度上已经成为当代建筑面临的重要挑战之一。现代主义曾许下救世主般的承诺,以拯救之名大规模推进拆建项目,以截然一"新"的建筑不断替代原有的"老"建筑。社会和建筑师们挣脱现代主义的桎梏,努力从这一原有的遗留问题中解放出来。无独有偶,20世纪90年代以来,建筑保护主义者也已经意识到,对文化遗产的继承如果仅限于修固建筑文物本身,也就是死气沉沉地见证过去的历史。[1]

就传承历史文化遗产这个理念本身而言,它体现出人类对自身价值的保护和对记忆的留存,这是人类寻求自我保护与发展的一个最根本的努力方向。如此看来,保护历史文化遗产是一种责任行为,它平等地涉及道德态度、规范政策和技术专长三个方面。可以把彰显"优秀普遍价值"[2]的文化产品开列一份目录,但是"文化遗产"绝不是一份目录就能够穷尽的,这主要因为文化本身熔铸在社区的日常生活和鲜活记忆的肌理之中,镶嵌混成,发展演变。

在这样的背景下,建筑就是一种特定的行为,不仅需要考虑周围环境的美学价值和文化价值,还需要考虑如何保护和提高人类生产的生态系统功能。因此,建筑既折射出一定的社会关系,同时也追溯着人类智慧积淀的历程。问题最终演变成如何追求建筑的"最佳实现",这有赖于建筑师们高度的个人觉悟感与集体责任意识。所有新的建筑提案都必须建立在过去荣耀的建筑本底之上,这种荣耀有时甚至会被夸大到言过其实的程度,成为神话,所以一旦我们将其称之为建筑,它就代表了一项艰苦卓绝的任务。

Building into the historic fabric has largely become one of the principal challenges of contemporary architecture. Unshackled from the messianic promises of Modernism and its large-scale programs of substitution of the "old" with a cleansed "new", societies and architects alike stand emancipated against their own built legacy. Since the 1990s, the same applies to the preservationist fixations that see heritage in the terms of artifacts maintained as lifeless testaments of their past[1].

In itself, the idea of heritage brings forth one of the most fundamental aspects in mankind's endeavor for self-preservation and evolution, namely to preserve human values and memory. In these terms, safeguarding heritage becomes an act of responsibility involving equally an ethical stance, normative policy, and technical expertise. Nevertheless "cultural heritage" is hardly exhausted as a mere catalog of artifacts that show "outstanding universal value"[2], mainly because culture itself is embedded and evolving in the fabric of everyday practices and the living memory of the community.

The act of building in such context becomes case-specific, taking into account the aesthetic and cultural valuations of the setting, as well as the preservation and empowerment of an ecosystem of human production. As such, it reflects both social bonds and the trace of an accumulated intellect. The problem ultimately becomes a call for "good practices", informed by an elevated awareness of our individual and collective responsibilities. Once we deem it "architecture", it becomes an almost Herculean task since all new proposals must stand their place from a glorified past that sometimes stretches to the point of myth.

## Architectural Insertions: Building into the Historic Fabric

*"Cities are dynamic organisms. There is not a single 'historic' city in the world that has retained its 'original' character: the concept is a moving target, destined to change with society itself."*[3]

In these few lines, all of our architectural preoccupations unfold into play, yet they unfold by bringing forth the idea of a living breathing ever-evolving continuity of human production. Cities stand as gradual accumulations of a collective intellect, invested with the potential for people to fulfill their aspirations as individuals as well as in society. This accumulation also acts as a vessel for human memory and values, carrying layers of the past into the future as a point of reference for further additions.

The problem space for such a context is that any propositions about what to do articulate in the present tense. Architects stand to confront issues that evade rigid definitions, thus seeking to establish firm foundations on slippery grounds. Should our trade had the benefit of undisputable solutions, building production of this sort would hardly raise any opposition. Then again, it would also hardly qualify as architecture, since what we fundamentally expect of it is to push the envelope further. Architects renegotiate their limits and boundaries, and offer propositions of artistic merit that ultimately have the potential to become part of the culture themselves. As such, the problem of architecture is to define the value

# Fabric

百代基金会_Pathé Foundation/Renzo Piano Building Workshop
伦敦政治经济学院学生中心_Student Center at the London School of Economics and Political Science/O'Donnell + Tuomey Architects
索纳·雷德大楼_Seona Reid Building/Steven Holl Architects
嵌入的建筑：融入历史肌理_Architectural Insertions: Building into the Historic Fabric/Angelos Psilopoulos

**嵌入的建筑：融入历史肌理**

"城市是动态的有机体。世界上没有任何一座'历史名城'能够保持'最初'特质；'这个理念'是一个变动不居的目标，它注定随着社会的发展变化而改变。"[3]

寥寥数语，将我们建筑行业所有的当务之急展露无疑，展示出一个鲜活生动的、不断发展演变的人类生产连续性的理念。矗立的城市见证了集体智慧的日积月累，蕴含着潜质，倾注了人们实现个人梦想以及在社会中有所作为的热切期望。这种积淀是保存人类记忆与价值观念的文脉，传承着由过去到未来层层叠叠的历史，为将来的进一步扩建提供参考。

难就难在在如此环境内我们"现在"应当做什么。但是建筑师们所要应对的设计问题对此并不作任何明确的规定，所以他们只能是在不甚牢靠的场地上想方设法建立起稳固的基础。假如我们这个行业的优势就是提供无可置疑的解决方案，那么这类保守的建筑产品就不会遭致任何异议。话又说回来，毫无争议的产品也很难有资格称之为建筑，因为从本质上来讲，我们还是期待建筑类产品能够挑战常规，有所突破。建筑师们还是要突破限制和界限，拿出不仅具有一定艺术价值，而且具备潜在实力，能够最终铸就文化传奇的设计提案。因此，建筑的问题是如何判定哪些价值需要被保留下来，如何使建筑与周边环境之间协调一致、和谐共生，以及如何能够满足建筑周边社区繁荣兴旺、蓬勃发展的期望。

这就要求设计师们具备强烈的感受力。用苏珊·麦克唐纳的话来讲，"在承载着历史价值的环境中，最成功的新建筑设计，必然对该环境的独有特色和专属特质有一个整体的认识，然后进行有针对性的设计。对于建筑文化保护工作而言，至关重要的是需要对该地区重要的历史文化意义的全面了解。……另外，建筑设计工作需要具体情况具体分析。建造一处核心建筑风格统一的城市中心，可能需要采用不止一种建筑设计表现途径，才能够实现丰富多样的建筑形态、规模以及表现特征。而在传承了传统的建筑工艺、建筑技法和建筑质料的城市的居民区，把这些传统风格发扬光大、延续城市历史文脉，对于建筑设计而言就显得尤为重要。"[4]

如何在同质性、历史的延续性、破除传统、和谐共生、并置或融合之间做出正确的选择，引发了无数激烈的争论，是该做出评判的时候了。公平地说——事实上专家咨询机构就这一问题[5]达成了一致意见——受强烈的主观喜好因素影响，人们的评判终究摇摆不定，无法形成普遍共识；在这样的背景下，建筑工作就成为一项极具挑战性的任务。如果我们还需要考虑来自以下两方面的压力，这种挑战性就更为明显了：(1)项目必须依赖一定的经济利益支撑；(2)项目必须建立在一个荣耀非凡的、甚至成为传奇神话的原有建筑本底之上。最后的挑战来自建筑本身，当代建筑作品总是试图突破建筑等级结构，渴望成为引人瞩目的轰动事件，成为带有醒目符号意义的地标性建筑，而忽视了建筑本应具有的深远价值。[6]

我们的探究总是具有两重性：连贯性与奇异性、环境与对比、传统与创新，似乎所有当代的建筑命题都必然要在这个二元对立的谱系范围内对应一个位置。好在在当前的争论中，建筑本身即作为明证，展示自己作为一个有价值的嵌入结构，如何融入集体记录，且为其正确性提供充分的论据。本文探讨的建筑就是这一论点的显著例证。

百代电影基金会是一栋位于百代老电影院旧址之上的建筑，地处巴黎的一个城市街区的内部庭院中。该建筑利用一系列建筑构架的支撑达到一种新的平衡，为原先密集的空间提供了良好的通风、采光和自由空间。在尽可能地满足自身功能需求的同时，也为毗邻的建筑提供了一些颇有质感的要素，例如，缓解所在区域的密集感，重新连接宽绰的内部视野。从街上看，该建筑掩映在其他建筑的"内部"，比有形形态更具无形张力，比物质实体更具本质气息。建筑的正立面是罗丹早年创作的装饰性雕刻文物，门后方伦佐·皮亚诺建筑工作室的杰作赋予了这层石质外观新的生命与活力。这是摆脱了外形束缚的建筑的得天独厚的优势，当然，同时还需要辅以必要的技术和驾驭抽象概念的敏锐洞察力。

伦敦政治经济学院的学生中心，主要解决的艰巨任务是要在一个密集的、承载历史价值的建筑场址上开发空间。从功能角度来讲，这是一栋多功能型建筑，必须容纳数量庞大的各类设施，以满足学生的日常学习生活所需。建筑场所本身受到周边建筑线、狭窄的视野、有限的空间，以及风格多元的建筑参照物等诸多限制。为应对这一挑战，奥唐

that is to be preserved, to achieve terms of coherence or suitability with its surroundings, and to allow for the aspirations of the community attached to it to flourish and thrive.

This calls primarily for heightened sensitivity. In the words of Susan Macdonald, *"Most successful new buildings designed in a valued historic context inevitably rely on an understanding of, and then response to, the special character and qualities of the context. As with any conservation work, understanding significance of the place is crucial. Also (…) it is case specific. A city center with an architecturally unified city core may need a different approach than one that has a variety of architectural forms, scales, and expressions. In an urban settlement that continues to sustain traditional craft and building techniques and materials, it may be extremely important to promote the continuation of these practices."*[4]

Come the time to pass judgment, choosing between homogeneity, historical continuity, break with tradition, symbiosis, juxtaposition or integration, becomes a point of fiery discourse. It is fair to say – and in fact expert consulting bodies on the subject[5] agree – that there is a strong element of subjective taste that sways final judgment away from universal consensus; the act of building in such a context consequently becomes a very challenging task. This is even more evident once we take into account the pressure that a project must sustain (a) from vested financial interest and (b) from a glorified – and even mythically charged – past. The final

1. "Global Strategy", <UNESCO World Heritage Center>, <http://whc.unesco.org>, (23 September, 2014).
2. "The Criteria for Selection". <UNESCO World Heritage Center>, <http://whc.unesco.org>, (15 October 2014).
3. "New Life for Historic Cities: The Historic Urban Landscape Approach Explained", United Nations Educational, Scientific and Cultural Organization, 2013, p.24, <http://whc.unesco.org/document/123570>, (3 January 2015).
4. Susan Macdonald, "Contemporary Architecture in Historic Urban Environments", Conservation Perspectives, The GCI Newsletter; Newsletter 26 Febuary 2011, <http://www.getty.edu/conservation/publications_resources/newsletters/26_2/contemporary.html>, (3 January 2015).
5. Such as English Heritage and CABE; See "The BiC Toolkit", <Building In Context>, <http://www.building-in-context.org/the-bic-toolkit/>, (accessed 3 January 2015).
6. Charles Jencks, "The Iconic Building Is Here to Stay", City, Vol.10, 2006, p.3~20, <http://dx.doi.org/10.1080/13604810600594605>.

奈－图米建筑师事务所制定了两条重要设计策略：利用缜密细致的折叠及其展开形成的空间，最大限度地发挥潜力效用；创造性地使用建筑材料，与原有环境的建筑材质特点形成呼应。采光设计按照视线的角度量身定制，既能够从街角的视角进行观赏，还要实现室内与户外的视觉交互。通过一系列的精心和扬长避短的设计，这栋建筑最终以其设计优势成功突破了可用空间狭小的局限性。

史蒂芬·霍尔建筑师事务所设计的索纳·雷德大楼，旨在成为与1909年查尔斯·雷尼·麦金托什设计的格拉斯哥艺术学院主楼"风格迥异而又互补"的建筑，后者是建筑史上的杰作，当之无愧的建筑典范。新建筑与产生文化认同的老建筑比邻而立，迫切期望能够确立自己的地位。这种并置引发新老建筑之间激烈的对话[7]，有时甚至令人忽略了这座新建筑正在取代"寒酸"的野兽派前辈的事实[8]。新建筑在材料和外形方面的使用几乎都与麦金托什的设计语言形成抗衡。史蒂芬·霍尔建筑师事务所的合作人詹姆斯·麦卡沃伊把建筑内部的管状通道称为"驱光空间"。楼内厚重的混凝土框架被"驱光空间"自上而下切割贯通，呈现出横截面效果。这些通道融功能性和空间性为一体，营造出诗意的空间效果。连续延展的银河般的绿色玻璃外墙"托起"格拉斯哥天空洒下的日光，以其轻盈亮丽的风格与麦金托什的主楼庄严凝重的外观形成反衬，而室内的混凝土结构也与1909年建筑采用的轻快纤细的框架结构相互制衡。史蒂芬·霍尔在接受奥利弗·温赖特采访时说："我们把麦金托什的所有做法都反其道而行之，这样一来设计就变得很容易了"[9]。

过去的历史不会封冻在时间的长河中，新建筑设计必须在原有的环境条件下满足所有的功能和需求，本章涉及的建筑杰作就是很好的证明。这些例证同时也揭示出当代建筑的困境——应该如何定位才能更好地走向未来，成为历史。苏珊·麦克唐纳为我们呈现了建筑设计的两条普遍路径："近年来，针对现代主义风格的介入，有些建筑师选择继续设计更具历史风貌的建筑，但是仍然采用现代的建筑材料和技术。其他建筑师则拒绝复古主义，主张每一代人都要展现本时代的风范。新的建筑应当体现出各个时代的建筑理念、技术、材料和建筑语言。拼贴模仿是建筑设计的禁忌。"

drop in the bucket comes from the aspiring architecture itself, as the contemporary break with the hierarchy of building substitutes long-term value with attention-grabbing events and soothing iconography with stimulating icons[6].

As always, our inquiry becomes invested with dualities: coherence vs. singularity, context vs. contrast, heritage vs. novelty. Given the circumstances, it seems that any contemporary architectural proposition necessarily entails a position within this spectrum. Better yet, it acts in its right as a statement of intention towards an ongoing debate, making a case for its very validity as a worthy addition to the collective memory. The buildings we examine in this article, are evident examples of this argument.

The Pathé Foundation is a building that takes the place of Pathé's old theater built in the interior courtyard of a Parisian city block. In a series of fundamental architectural gestures it reinstates a certain balance, allowing for the air, light and free space to enter in the previously dense space. As much as it accommodates its own functional program it also offers its neighbors with elemental qualities such as the relief of density within the area that it occupies or the generous re-articulation of interior views. Seen from the street, it looks as if it emerges "from within", more immaterial than material, more essence than substance. Behind the heritage facade sculpted by a young Rodin, RPBW offer a new pulsating heart to breathe life into the stone shell. This is the privilege of an architecture unshackled by form, and, at the same time, invested with the necessary technology and insight to harness the abstract.

The Student Center at the London School of Economics and Political Science tackles, first and foremost, the difficult task to develop space in a dense, as well as historically charged, site. By program, it is a multi-functional building which must incorporate a significant number of facilities accommodating the necessities of everyday student life. The site itself is restricted to the surrounding building lines, narrow views, limited space, and diverse stylistic references. In response to the challenge, O'Donnell+Tuomey Architects bring forth two significant design strategies: the thoughtful folding and subsequent unfolding of space in order to maximize its potential, and the creative use of materials answering interpretatively to the character of the existing settlement. The result is tailored to lines of sight, to be viewed from street corner perspectives and to make visual connections between internal and external circulation. It is a building perceived through a series of carefully choreographed vantage points, transcending the limitations of the available space.

The Seona Reid building by Stephen Hall Architects aims to stand, by its own statement, as a "complementary contrast" to Charles Rennie Mackintosh's 1909 Glasgow School of Art building – if anything a reference masterpiece in the History of Architecture. As such, the new building looks, first and foremost, to establish its own presence against a culturally sanctioned counterpart. This confident juxtaposition has ignited fiery discourse[7], sometimes forgetting that the new building is taking the place of a "rather shabby" brutalist predecessor[8]. The building uses materials and

7. Rory Olcayto, "Holl's Glasgow Art School Building: First Reaction", Architects' Journal, 2014, <http://www.architectsjournal.co.uk/comment/holls-glasgow-art-school-building-first-reaction/8658938.article>, (5 January 2015).
8. Douglas Murphy, "Reid Building by Steven Holl", Icon Magazine, 2014, <http://www.iconeye.com/architecture/news/item/10201-reid-building-by-steven-holl>, (5 January 2015).
9. Oliver Wainwright, "Green Giant: Can Glasgow's New School of Art Eclipse Mackintosh's Marvel?", The Guardian, 2014, <http://www.theguardian.com/artanddesign/2014/mar/01/charles-rennie-mackintosh-glasgow-school-art>, (7 January 2015).
10. "The art of inserting a new building into a historic city block means engaging in an open, physical dialogue with the existing city buildings. Building onto a structure also presents an opportunity for a wide-ranging renovation project, a reclaiming of space.", 'Pathé Foundation', Renzo Piano Building Workshop, <http://www.rpbw.com/project/81/pathe-foundation/>, (5 January 2015).

　　无论选择哪一种设计途径，当代建筑都面临着自身的存在性这个关键性问题，其次还需要面对自己的虚荣心问题。查尔斯·詹克斯的作品展示了一个建筑等级秩序被"神秘符号"打乱的时代，神秘符号指的是与丰碑式建筑物形成鲜明对比的一种象征性建筑物，这类建筑的自信之处在于，必将打破象征主义建筑的等级层次，发展成为一个和谐连贯的整体。詹克斯引用切斯特顿的话指出，问题在于"当人们不再信仰上帝，他们不是什么也不信仰，而是信仰一切"。如今，任何建筑都有权利希望在建筑史上扮演重要角色，哪怕是那些需要回望过去，借鉴其他建筑风格和证实自身有效性的建筑。

　　人们想要保留的究竟是什么？这个问题有待建筑师们来回答。然而，建筑还在一刻不停持续增建，信心满满地应对复杂多元的背景带来的挑战。如果我们分析伦佐·皮亚诺10的设计就会发现，他所设计的项目都与建筑修复与保护问题相关，都是对空间的回收再利用——也就是说，不仅仅是对建筑遗产进行维护，更要在现存的条件下发展和活化这些文化遗产。皮亚诺设计的百代电影基金会大楼，恰如其分地施展了他的建筑理念，彰显出卓越非凡的建筑品质。该建筑是对城市的记忆、审美品质、地域特色的绝好证明。

　　本章开篇已经讲到，我们需要对历史文化的意义和价值具有非常敏锐的感知能力。同时，一旦建筑师本人想要凸显自己的重要地位，在集体记忆的积累和知识产权方面引人注目，虚荣心就会造成巨大危害。对此，环境保护专家能够发挥重要的咨询作用，但是道德义务仍需建筑本身来承担。再次引用苏珊·麦克唐纳的话来结束本章，"建筑设计师们要负起责任，……确保自己的工作不仅不会破坏已有的建筑环境，反而能够锦上添花。"以下我们将要考察的项目正是对这个论点的有力支撑。

form in an almost per-case counterbalance to Mackintosh's design language. A heavy concrete shell is cut cross-section by what SHA's partner James McAvoy deems as "driven voids of light" – tubular voids that intersect function and space with a poetic in-between. A continuous surface of milky green glass "picking up the light" from the sky of Glasgow seeks to establish lightness against Mackintosh's gravitas, while the concrete interior counterbalances the airy and slender framework of the 1909 building. As Stephen Holl puts it in an interview with Oliver Wainwright, *"We made it easy by doing the opposite of everything Mackintosh did"*9.

All of these efforts stand in testament that the past is hardly frozen in time, and that functions and needs must still be serviced in existing settlements. They also reveal the agony of the present to establish its place towards a future past. In her assessment, Susan Macdonald presents us with two general approaches to design:

*"In recent times, in reaction to modern interventions, some architects have chosen to continue to design buildings in a more historical style while nevertheless utilizing modern materials and technologies. Others abhor historicism and argue that each generation should represent its own time. New layers should represent the ideas, technology, materials, and architectural language of each generation. Pastiche is a dirty word."*

In both cases, contemporary architecture is critically confronted with its own presence. Subsequently it is also confronted with its own vanity. Charles Jencks showcases an era where the hierarchical order of building has been disrupted by "enigmatic signifiers", referring to a-symbolic construction in stark contrast to the monument and its confident place in a hierarchy of symbolism necessarily amounting to a coherent whole. The problem is, as he states quoting Chesterton, that "when people stop believing in God, they don't believe in nothing - they believe in anything". Today any building can aspire to assume that role, including the ones that look back to the past for stylistic reference and validity.

What do people want to hold on to? The question is open to architects. Nevertheless, architecture keeps expanding its own agenda, standing confident against the challenges that are posed by a multi-faceted context. Should we paraphrase Renzo Piano[10], these projects are at once a problem of renovation and conservation, and a reclaiming of space – that is, not only safeguarding built heritage but also expanding it in living, breathing conditions. The foundation, upon which such an approach is possible, is the very substance of an exceptional quality. It may manifest in terms of memory, esthetic quality, or a sense of place.

As we stated in the beginning, this requires an acute sensibility on meaning and value. At the same time, it reveals the detrimental dimension of vanity, once the architect claims his place in the collective accumulation of memory and intellectual property. In this endeavor, conservation specialists may play a critical advisory role, but the ethical burden stays with architecture. Again in the words of Susan Macdonald, it remains *"The responsibility of designers (…) to ensure that their work contributes to and enriches rather than diminishes the built environment"*. The projects we examine in the following pages establish this argument in the present tense.

Angelos Psilopoulos

# 百代基金会

Renzo Piano Building Workshop

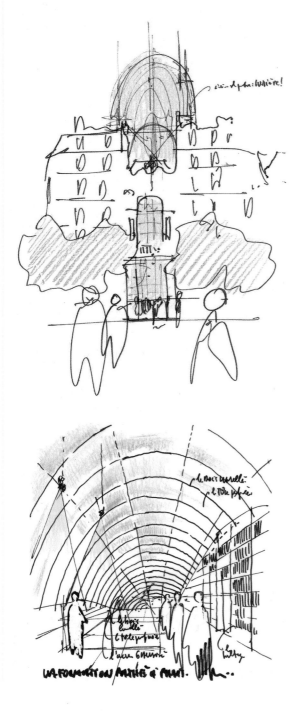

将一座现代建筑嵌入一处历史名城街区,这种艺术形式意味着需要与当地的已有建筑开启一段开放、实质性的对话。在原有建筑基础之上建造新的建筑,也代表了改造项目(即改造一片区域)的广泛普及性。

百代电影遗产保护组织(百代基金会)的新总部是一个出乎意料的存在,从它所处的庭院中间位置一眼望去,看到的是一个巨大的曲形体量,固定在几处支撑点上。地面上,一片白桦树林形成了这一带密集街区里的一座鸟语花香的"岛屿"。

百代基金会致力于保存"百代电影公司"的遗产并推广其摄影技术。它的新总部坐落于法国第十三行政区一处街区的中心位置,这里还有一间由19世纪中期的古老剧院改建的、又在19世纪60年代再次崛起的电影院(巴黎最早成立的电影院之一)。

新建筑将设有档案管理室以及一些用于展示临时及永久珍藏品的展览室,其中包括一间70座的放映厅,另外还有行政办公区。

为了更好地打破场所的限制,项目计划拆除两座原有建筑来创造一个有机"生物"。这样的设想是为了响应基金会对功能性及代表性规划的需求,同时提升新建筑周围空间的环境品质。

位于高伯兰大街的外立面,由于其具有的重要历史和艺术价值,将得到修复及保护。外立面被艺术家奥古斯特·罗丹初期的雕塑作品所装饰,不仅是历史性的地标,也是高伯兰区的象征性建筑。

一座崭新的透明建筑矗立在外立面的后面,作为基金会的公共入口。项目的主要功能区集中于中央院内的第二座建筑,透过其一层的透明外壳,能够欣赏到室内花园景象,使建筑看上去像是一座温室。

建筑的独特设计主要是由场所的限制和要求所决定的。其与周边建筑保持适当距离以示尊重,确保了相邻建筑的自然采光及通风。该项目通过减少面积,在场所后身的空地上创造了一座花园。

建筑的上部是由玻璃制成的,为基金会的办公区提供自然采光。

从白天的街区方向望去,这座蛋形建筑看上去只是一座立面经过修复的普通建筑,低调而不起眼,然而当夜幕降临,灯光就会从透明外壳中摇曳而出,展示它独有的魅力。

## Pathé Foundation

The art of inserting a building into an historical city block means engaging in an open, physical dialogue with those already there. Building onto an extant structure also presents an opportunity for a more widespread renovation project, a reclaiming of space. The new headquarters of the Fondation Jérôme Seydoux-Pathé is an unexpected presence, a curved volume one glimpses floating in the middle of the courtyard in which it sits, anchored on just a few supports. On the ground, there is a stand of birch trees, a floral island set in the dense mineral context of the city.

The Fondation Jérôme Seydoux-Pathé is an organization dedicated to the preservation of Pathé's heritage, and to the promotion of the cinematographic art. Its new headquarters sits at the center of a block in the XIII Arrondissement, where an old mid-19th century theater- transformed into a cinema(one of the first ones in Paris) in the mid-1900s and then radically transformed again in the 1960s-once stood.

鸟瞰图_新建筑  aerial view_new building　　　　　鸟瞰图_老建筑  aerial view_existing building

项目名称：Pathé Foundation
地点：73, Avenue des Gobelins, 75013 Paris, France
建筑师：Renzo Piano Building Workshop
主要合作者：B. Plattner, T. Sahlmann
合作者：G.Bianchi / Model A. Pachiaudi, S. Becchi, T. Kamp, S. Moreau, E. Ntourlias, O. Aubert, C. Colson, Y. Kyrkos
结构工程师：VP Green
MEP：Inex / 3D模型制作：Arnold Walz / 造价顾问：Sletec
甲方：Fondation Jérôme Seydoux-Pathé
可持续性：Tribu
音效工程师：Peutz
照明：Cosil / 室内：Leo Berellini Architecte
总建筑面积：2,200m² / 场地面积：839m² / 高度：25m
功能：design and construction of building housing the Fondation Seydoux-Pathé headquarters, the archives, permanent exhibition space for the cinematographic collection of Charles Pathé, 70-seat screening room, offices
竣工时间：2014
摄影师：©Michel Denancé(except as noted)

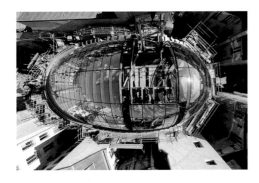

西南立面 south - west elevation

The new building will house Pathé's archives, some spaces for temporary exhibitions as well as for the permanent collections including a 70-seat screening room, and the offices of the Foundation.

The project called for the demolition of the two existing buildings to create an organic "creature" that better responds to the restrictions of the site. The idea was to respond to the functional and representative program requested by the Fondation, while at the same time increasing the quality of the space surrounding the new building. The facade on the Avenue des Gobelins has been restored and preserved, due to its historical and artistic value. Decorated with sculptures by a young Auguste Rodin, it is not only a historical landmark, but also an iconic building for the Gobelins area.

A new transparent building just behind the facade functions as the foundation's public access. Looking like a greenhouse, it offers a view on the interior garden through the transparent ground floor of a second building in the central court that houses the project's main functions.

The peculiar design of this building is determined by the site's major limits and requirements. While respecting the distances with the surrounding buildings, the building improves the neighbour's access to natural light and air. By reducing the footprint, the project creates space for a garden in the back of the site.

The upper part of the building is made of glass, providing natural light for the office spaces of the Fondation.

From the street the building is only perceived through the restored facade like a discreet presence during the daytime, while it will be softly glowing at night.

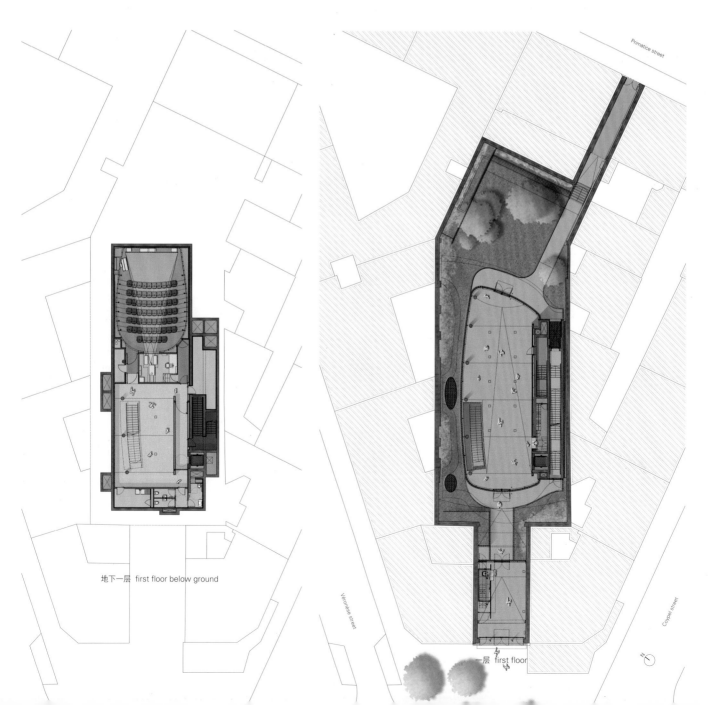

地下一层 first floor below ground

一层 first floor

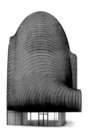

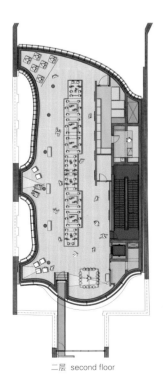
二层 second floor

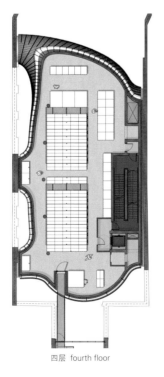
四层 fourth floor

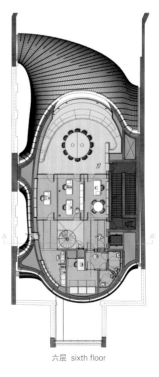
六层 sixth floor

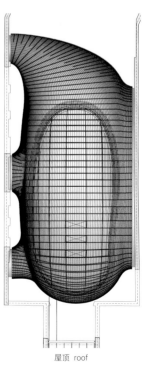
屋顶 roof

照片提供：© Paul Raftery (courtesy of the architect)

透光情况 perforation degree

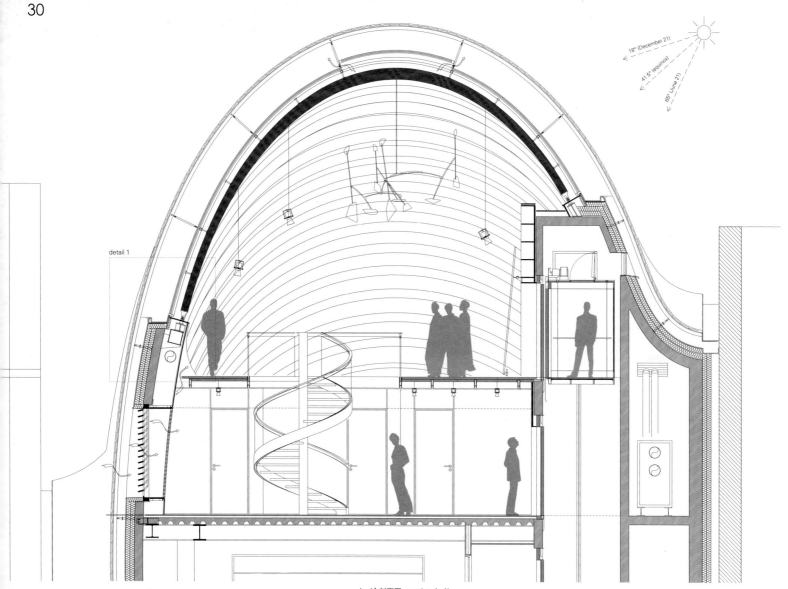

A-A' 剖面图  section A-A'

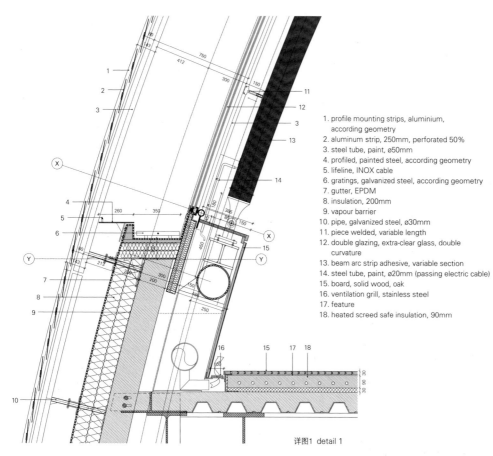

1. profile mounting strips, aluminium, according geometry
2. aluminum strip, 250mm, perforated 50%
3. steel tube, paint, ø50mm
4. profiled, painted steel, according geometry
5. lifeline, INOX cable
6. gratings, galvanized steel, according geometry
7. gutter, EPDM
8. insulation, 200mm
9. vapour barrier
10. pipe, galvanized steel, ø30mm
11. piece welded, variable length
12. double glazing, extra-clear glass, double curvature
13. beam arc strip adhesive, variable section
14. steel tube, paint, ø20mm (passing electric cable)
15. board, solid wood, oak
16. ventilation grill, stainless steel
17. feature
18. heated screed safe insulation, 90mm

详图1  detail 1

# 伦敦政治经济学院学生中心

O'Donnell+Tuomey Architects

项目设计将学生的各项设施融为一体,位于一个屋顶之下。这栋多功能建筑包括会议室、酒吧、咖啡厅、多媒体设施、祈祷室、办公室、健身房、就业中心、舞蹈室和社交空间。

项目地处斯特兰德保护街区,周边环境复杂,且场地被周边的建筑控制线所限制。威斯敏斯特理事会的规划师对各项技术指标严格把关,旨在使现代设计与周围的环境背景相得益彰。在整个承建过程中,规划师们坚持追求以精心的施工来打造恒久的建筑品质。

在伦敦政治经济学院的城市中心校区,街道狭窄是该校区的一大显著特点,项目就坐落在狭窄街道的交汇点上。其多面的立面对建筑的轻质外围护结构进行处理,并量身定制人们的视野,既能使人们从街角的视角进行观赏,又实现了室内与户外流线的视觉连接。砖墙表皮沿建筑折线切断,形成大型玻璃区,将视野框入其中。对周边环境特点的分析成为影响特定场所建筑设计的基本原则的要素。

设计还考虑到现代学生中心的动态特征。场地复杂的几何外观决定了建筑内部各楼板形状的不规则性,经过合理的布局,使形状各异的楼板各自发挥独特的功能。空间在平面与剖面方向自由流动,楼梯盘旋回转,在每个楼层空间形成会聚平台。

伦敦是砖造建筑之城。该建筑外部是红砖结构,砖体之间错落有致,形成敞开式工作格局。白天,日光在室内投下斑驳的光影,夜晚,室

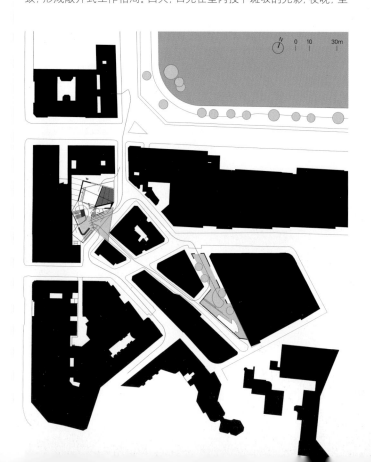

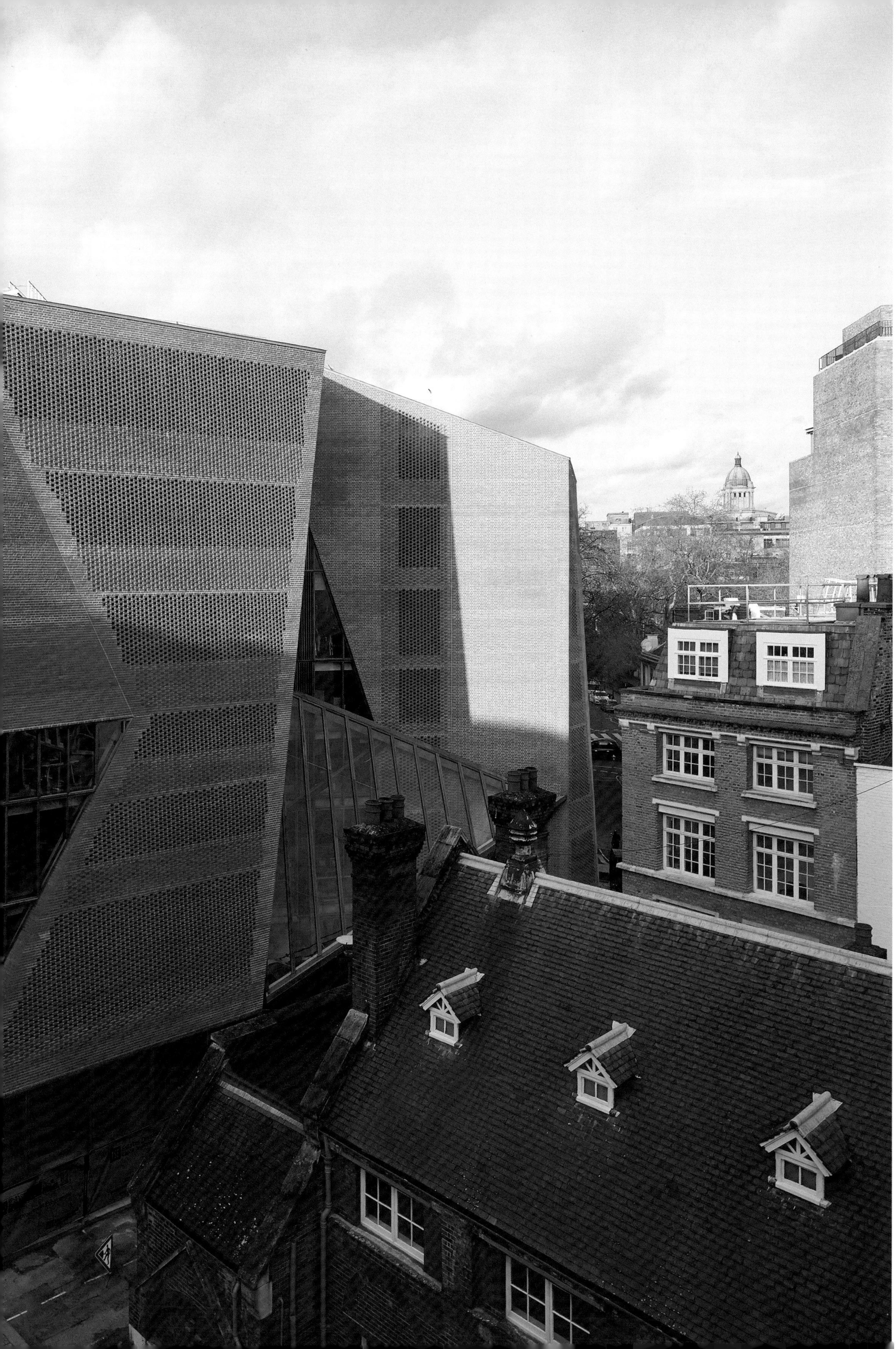

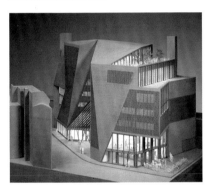

内的灯光将整栋建筑打造成一盏发光的格子灯笼。建筑内部铺设坚固的木质地板，如同一个宜居的大仓库，具备很强的功能适应性。大楼采用钢筋混凝土与钢结构混合建造模式，在空旷的室内上部空间搭建钢桁架或肋形混凝土板，起到了很好的支撑作用。圆形钢柱支撑着横跨大型体量的办公楼层，并且突出了咖啡厅的开放平面。混凝土天花板能够起到蓄热作用，而悬浮的声反射云可以使室内的声音变得更加柔和。楼内走廊采用开放式设计，每条走廊至少在一个方向上可见日光，并且有景致可供观赏。每间办公室的工作区都能看到户外的街区风景。天窗的设计满足了地下室活动场所的采光要求。

建筑以无障碍和包容性设计作为设计规划的重点考虑因素。入口处不设台阶，楼板的设置宽敞平坦，无台阶。环形的开放型通道配以清晰的路径引导标志，公共服务设施置在各楼层的同一位置。建筑中央精心设计了宽敞的楼梯，设计完全符合授权的检查机构认可的标准及细节要求。

### Student Center at the London School of Economics and Political Science

The brief was to bring student facilities together under one roof. The multi-functional building includes a venue, pub, cafe, media, prayer, offices, gym, careers, dance studio and social spaces.

The site lies within the Strand Conservation Area. The context was complex and the site was restricted by surrounding building lines. Specifications were closely monitored by Westminster planners, who supported the ambition for a contemporary design integrated with its setting. Throughout the building process, the planners maintained a commitment to the enduring quality of carefully crafted construction.

The site is located at the knuckle-point convergence of narrow streets that characterise the LSE city centre campus. The faceted facade operates with respect to the Light Envelope and is tailored

- metal roof finish
- untreated hardwood
- flemish bond brickwork
- perforated flemish bond brickwork with glazing behind
- textured flemish bond brickwork
- steel doors
- metal capping
- solar panels

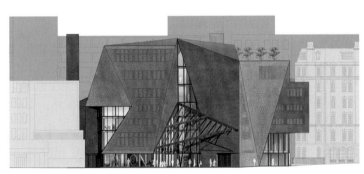

东立面 east elevation

北立面 north elevation

西立面 west elevation

南立面 south elevation

砌砖节点详图
brick junction detail

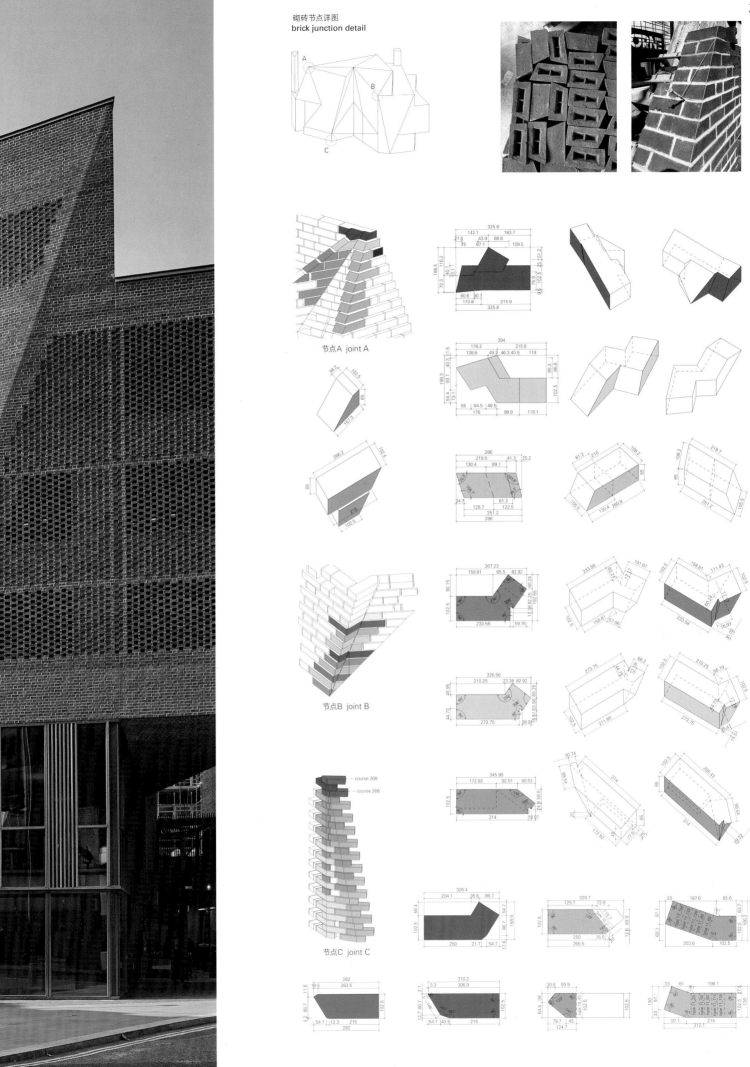

节点A joint A

节点B joint B

节点C joint C

项目名称：Saw Swee Hock Student Centre, London School of Economics
地点：London, UK
建筑师：O'Donnell+Tuomey Architects
主管：John Tuomey, Sheila O'Donnell
合伙人：Willie Carey / 项目建筑师：Geoff Brouder
项目经理：Turner & Townsend
项目团队：Laura Harty, Kirstie Smeaton, Gary Watkin, Anne-Louise Duignan, Ciara Reddy, Jitka Leonard, Iseult O'Cleary, Henrik Wolterstorff, Mark Grehan, Monika Hinz
执行建筑师：O'Donnell+Tuomey Architects
结构工程师：Dewhurst Macfarlane and Partners / Horganlynch Consulting Engineers
服务和环境工程师：BDSP
安保、消防、音效、交通以及物流工程：Arup
餐饮：Tricon Foodservice Consultants / 通道：David Bonnett Associates / 考古：Gifford
工料测量师：Northcroft / 规划顾问：Turley Associates / 界墙设计顾问：Anstey Horne
建筑控制顾问：Carillion / CDM协调员：Gardiner & Theobald / 主要承包商（设计&建造）：Geoffrey Osborne Limited
甲方：London School of Economics and Political Science, Estates Division.
室内有效楼层面积：6,101m² / 造价：GBP 24,115,603 / 施工时间：2011.5——2013.12
摄影师：©Dennis Gilbert/VIEWpictures.co.uk

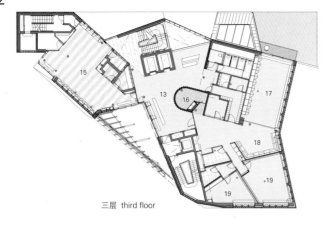
三层 third floor

七层 seventh floor

二层 second floor

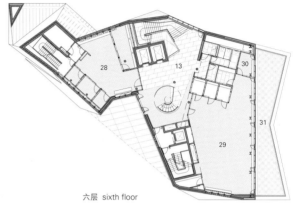
六层 sixth floor

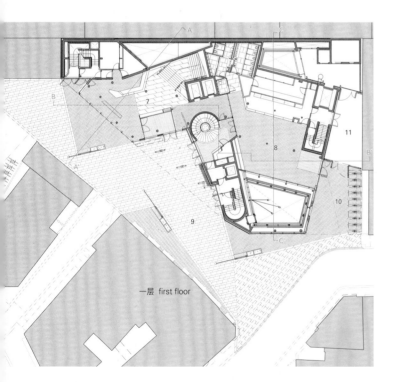
一层 first floor

| | |
|---|---|
| 1 | 车间 |
| 2 | 交流室 |
| 3 | 员工室 |
| 4 | 活动室 |
| 5 | 活动夹层 |
| 6 | 厨房 |
| 7 | 接待处 |
| 8 | 酒吧 |
| 9 | 人行道 |
| 10 | 自行车存放处 |
| 11 | 院子 |
| 12 | 活动中心 |
| 13 | 流线区 |
| 14 | 咖啡室 |
| 15 | 媒体中心 |
| 16 | 主教祈祷室 |
| 17 | 多信仰祈祷室 |
| 18 | 多信仰社交中心 |
| 19 | 祈祷室 |
| 20 | 伦敦经济学院住宅 |
| 21 | 等候室 |
| 22 | 办公室 |
| 23 | 存储室 |
| 24 | 会议室 |
| 25 | 体育馆 |
| 26 | 更衣室 |
| 27 | 淋浴室 |
| 28 | 职业中心 |
| 29 | 职业服务办公室 |
| 30 | 管理人员办公室 |
| 31 | 绿色屋顶 |
| 32 | 运动室 |
| 33 | 咖啡室/饮料吧 |
| 34 | 学生会办公室 |
| 35 | 露台 |

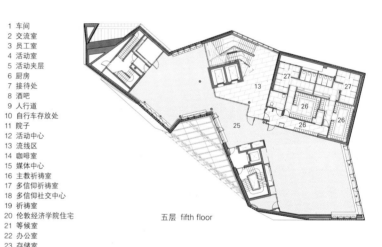
五层 fifth floor

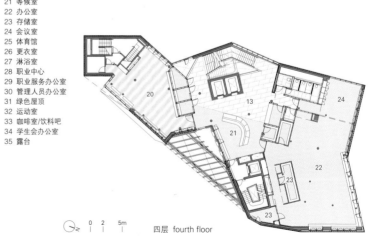
四层 fourth floor

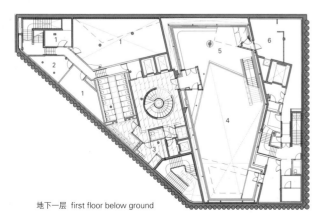
地下一层 first floor below ground

1. plant
2. communication room
3. staff room
4. events space
5. events mezzanine
6. kitchen
7. reception
8. pub
9. pedestrian street
10. bicycle store
11. yard
12. activity center
13. circulation area
14. cafe
15. media center
16. primate prayer room
17. multi faith prayer
18. inter faith social space
19. prayer room
20. LSE residences
21. waiting space
22. office
23. store
24. meeting room
25. gym
26. changing room
27. shower room
28. careers center
29. careers service room
30. managers office
31. green roof
32. exercise studio
33. coffee/juice bar
34. student union room
35. terrace

to lines of sight, to be viewed from street corner perspectives and to make visual connections between internal and external circulation. The brick skin is cut along fold lines to form large areas of glazing, framing views. Analysis of the context has influenced the first principles of a specific site's architectural design.

The building is designed to embody the dynamic character of a contemporary Student Centre. The complex geometries of the site provided a starting point for the arrangement of irregular floor plates, each particular to its function. Space flows freely in plan and section, with stairs turning to create meeting places at every level.

London is a city of bricks. The building is clad with bricks, with each brick offset from the next in an open work pattern, creating dappled daylight inside and glowing like a lattice lantern at night. The building has the robust adaptability of a lived-in warehouse, with solid wooden floors underfoot. The structure is a combination of reinforced concrete and steelwork. Steel trusses or ribbed concrete slabs span the big spaces. Circular steel columns prop office floors between the large span volumes and punctuate the open floor plan of the cafe. Concrete ceilings contribute thermal mass with acoustic clouds suspended to soften the sound. There are no closed-in corridors. Every corridor has daylight and views in at least one direction. Every office workspace has views to the outside world. The basement venue is daylight from clerestory windows.

The building is designed with accessibility and inclusive design as key considerations. Approaches are step free. Floor plates are flat without steps. Circulation routes are open and legible with clearly identifiable way-finding. Services are located at consistent locations. The central wide stair was carefully designed to comply with standards and details agreed with the approved inspector.

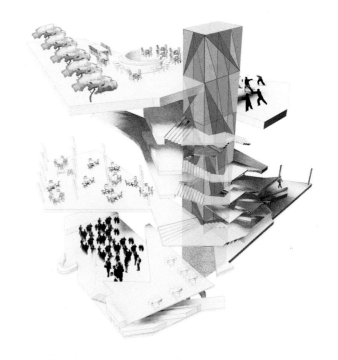

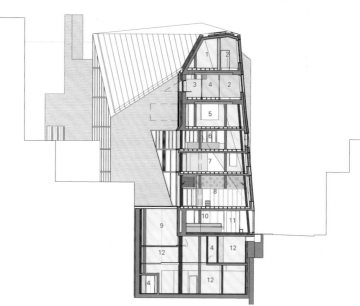

A-A' 剖面图 section A-A'

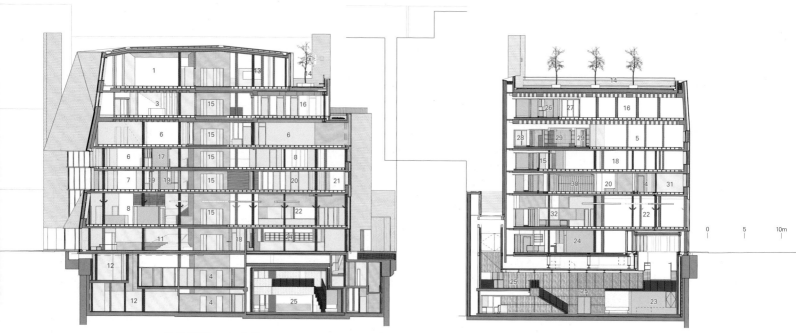

B-B' 剖面图 section B-B'

C-C' 剖面图 section C-C'

1 运动室
2 楼梯
3 职业中心
4 大厅
5 体育馆
6 伦敦经济学院住宅
7 媒体中心
8 活动中心
9 排烟通风室
10 接待处/安保处
11 接待处入口门厅
12 车间
13 咖啡室/饮料吧
14 屋顶花园
15 流线区
16 职业服务办公室
17 通知区
18 办公室
19 广播亭
20 多信仰社交中心
21 多信仰祈祷室
22 咖啡室
23 票务处
24 酒吧
25 活动室
26 茶室
27 咨询室
28 淋浴室
29 更衣室
30 储物柜
31 祈祷室
32 调查室

1. exercise studio
2. stair
3. careers center
4. lobby
5. gym
6. LSE residences
7. media center
8. activity center
9. smoke ventilation plenum
10. reception/security
11. reception entrance foyer
12. plant
13. coffee/juice bar
14. roof garden
15. circulation area
16. careers service room
17. advice
18. office
19. radio booth
20. inter faith social space
21. multi-faith prayer
22. cafe
23. ticket office
24. pub
25. events space
26. tea room
27. consultation room
28. shower room
29. changing room
30. lockers
31. prayer room
32. survey

城市建造——融入历史肌理 Urban How – Into the Historic Fabric

索纳·雷德大楼 Steven Holl Architects

索纳·雷德大楼与1909年查尔斯·雷尼·麦金托什设计的格拉斯哥艺术学院主建筑的风格迥异但相辅相成,两栋建筑和谐共生,相互提升彼此的整体品质。薄薄的且半透明的建筑材料与麦金托什大楼的砖石结构形成对比,充足的光线体量彰显了学院在城市肌理中的活力,展现出具有前瞻性的艺术生活。

项目独特的室内外设计加速了21世纪艺术学院新典范的树立。由内而外,设计融入了项目的功能需求与心理欲求;由外而内,项目连接了城市校园,并且与对面的麦金托什大楼产生联系,设计体现出学院热切渴望能够在城市肌理中发挥重要的作用。

麦金托什独具匠心的光线设计令人惊叹,启发建筑师对不同光线下的体量进行规划。工作室/研讨室是建筑中的基本体块。

空间定位不仅反映出彼此间相互依存的关系,也体现了不同空间对自然光线的不同需求。位于北面的工作室安装了大面积的朝北面倾斜的玻璃,以最大限度地接收理想的北面射来的高质量漫射光。餐厅和办公室等对自然光线质量没有同样要求的空间位于南面,它们利用遮阳设备调节日光的获取量,平衡住户需求与空间热性能之间的关系。

"驱光空间"实现了结构、空间调节和光线的融合。"驱光空间"的采光井为建筑提供自然光线,根据强度和天空色彩的不断变化,建立与外界之间的直接联系。此外,它们保证了整栋建筑的垂直流线,避免了对空调的依赖。

沿着南立面,与麦金托什设计的主工作室处于同一高度的沙质低地景观凉廊成为学院的室外社交核心,向城市开放。一些石质工艺品点缀的自然植被将水流引向一个小型回收池,斑驳的阳光透过水面反射到室内的天花板上。

贯穿整个新格拉斯哥艺术学院的"环路"鼓励不同的系所之间进行"创造性的碰撞",这些系所都是学院工作的重心。阶梯坡道形成的开放环路将大厅、展览空间、项目空间、阶梯教室、会议室、工作室、研讨室以及用于非正式聚会和展览的绿色露台等所有的主要空间连接起来。

## Seona Reid Building

The Seona Reid Building is in complementary contrast to Charles Rennie Mackintosh's 1909 Glasgow School of Art – forging a symbiotic relation in which each structure heightens the integral qualities of the other. A thin translucent materiality has considered contrast to the masonry of the Mackintosh building – volumes of light which express the school's activity in the urban fabric embodying a forward-looking life for the arts.

This project's unique interior and exterior forces on the design are the catalysts for creating a new 21st century model for the art school. Working simultaneously from the inside out – engaging the functional needs and psychological desires of the program, and the outside in – making connections to the city campus and relating to the Mackintosh building opposite, the design embodies the school's aspirations in the city's fabric.

Mackintosh's amazing manipulation of the building section for light in inventive ways has inspired the architects' approach towards a plan of volumes in different light. The studio/workshop is the basic building block of the building.

项目名称：Seona Reid Building
地点：Glasgow, United Kingdom
建筑师：Steven Holl Architects
项目团队：Design architects_Steven Holl, Chris McVoy
Partners in charge_Chris McVoy, Noah Yaffe / Project architect_Dominik Sigg
Assistant project architect_Dimitra Tsachrelia / Competition team_Dominik Sigg, Peter Adams, Rychiee Espinosa / Rychiee Espinosa, Scott Fredricks, JongSeo Lee, Jackie Luk, Fiorenza Matteoni, Ebbie Wisecarver
合作建筑师：JM Architects
项目经理：Turner & Townsend
总承包商：Sir Robert McAlpine / 工程师：Ove Arup & Partners
工料测量师：Turner & Townsend / CDM协调师：Cyril Sweett
景观建筑师：Michael Van Valkenburgh and Associates
规划师：Turley Associates / 甲方：The Glasgow School of Art
总建筑面积：11,250m²
竞标时间：2009 / 竣工时间：2014
摄影师：
©Iwan Baan(courtesy of the architect) - p.46~47, p.48, p.49, p.50~51, p.52, p.53, p.54
©Alan McAteer(courtesy of the architect) - p.57[bottom], p.59

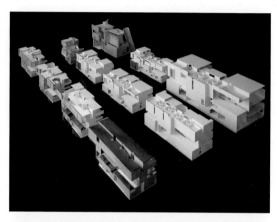

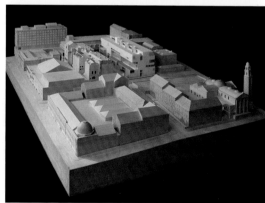

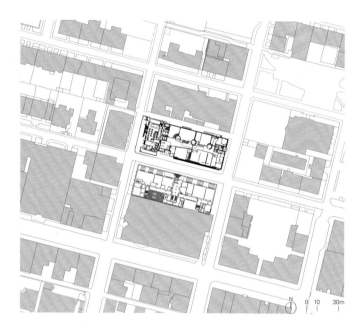

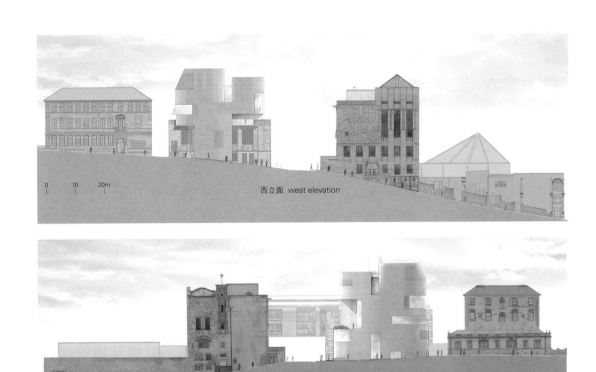

西立面 west elevation

东立面 east elevation

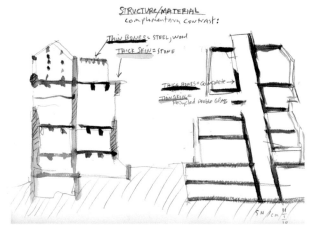

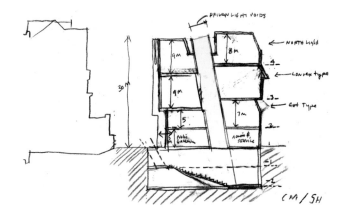

Spaces have been located not only to reflect their interdependent relationships but also their varying needs for natural light. Studios are positioned on the north facade with large inclined north facing glazing to maximize access to the desirable high quality diffuse north light. Spaces that do not have a requirement for the same quality of natural light, such as the refectory and offices, are located on the south facade where access to sunlight can be balanced with the occupants needs and the thermal performance of the space through application of shading.

"Driven voids of light" allow for the integration of structure, spatial modulation and light. The "Driven Void's" light shafts deliver natural light providing direct connectivity with the outside world through the changing intensity and color of the sky. In addition, they provide vertical circulation through the building, eliminating the need for air conditioning.

Along the south elevation, at the same height as the Mackintosh main studios, a landscape loggia in the form of a Machair gives the school an exterior social core open to the city. The natural vegetation with some stone work routes the water into a small recycling water pond which will reflect dappled sunlight onto the ceiling inside.

A "Circuit of Connection" throughout the new GSA encourages the "creative abrasion" across and between departments that are central to the workings of the school. The open circuit of stepped ramps links all major spaces – lobby, exhibition space, project spaces, lecture theater, seminar rooms, studios, workshops and green terraces for informal gatherings and exhibitions.

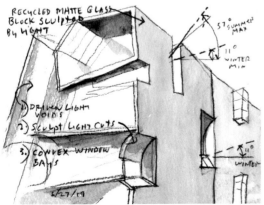

可持续性示意图 sustainability diagram　　　　　　　　　季节性太阳高度示意图 seasonal sun altitude diagram

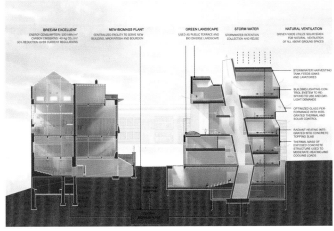
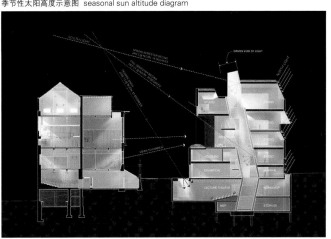

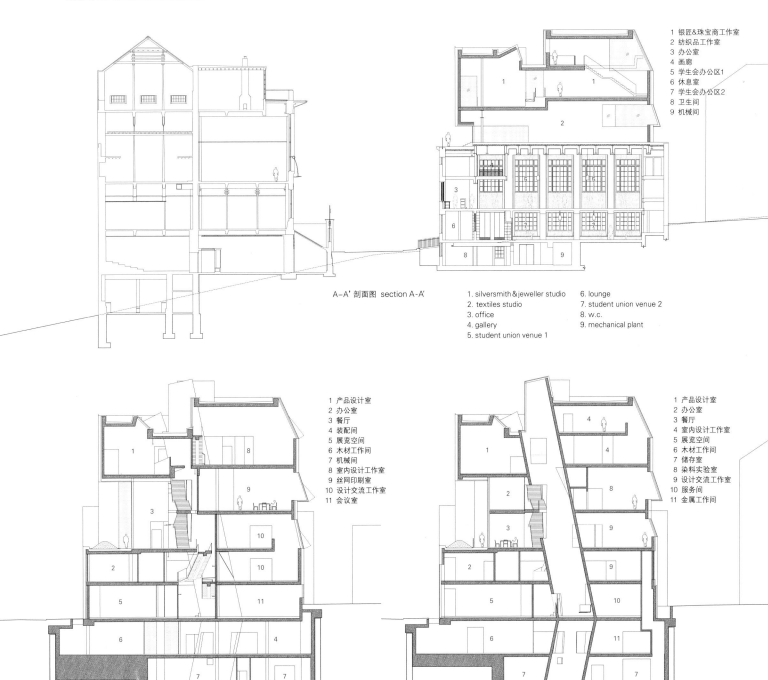

A-A' 剖面图 section A-A'

1 银匠&珠宝商工作室
2 纺织品工作室
3 办公室
4 画廊
5 学生会办公区1
6 休息室
7 学生会办公区2
8 卫生间
9 机械间

1. silversmith & jeweller studio　　6. lounge
2. textiles studio　　　　　　　　7. student union venue 2
3. office　　　　　　　　　　　　8. w.c.
4. gallery　　　　　　　　　　　　9. mechanical plant
5. student union venue 1

B-B' 剖面图 section B-B'

1 产品设计室
2 办公室
3 餐厅
4 装配间
5 展览空间
6 木材工作间
7 机械间
8 室内设计工作室
9 丝网印刷室
10 设计交流工作室
11 会议室

1. product design studio　　5. exhibition space　　9. screen printing studio
2. office　　　　　　　　　　6. wood workshop　　　10. communication design studio
3. refectory　　　　　　　　7. mechanical plant　　11. seminar room
4. assembly spaces　　　　　8. interior design studio

C-C' 剖面图 section C-C'

1 产品设计室
2 办公室
3 餐厅
4 室内设计工作室
5 展览空间
6 木材工作间
7 储存室
8 染料实验室
9 设计交流工作室
10 服务间
11 金属工作间

1. product design studio　　5. exhibition space　　9. communication design studio
2. office　　　　　　　　　　6. wood workshop　　　10. service
3. refectory　　　　　　　　7. storage　　　　　　11. metal workshop
4. interior design studios　　8. dye lab

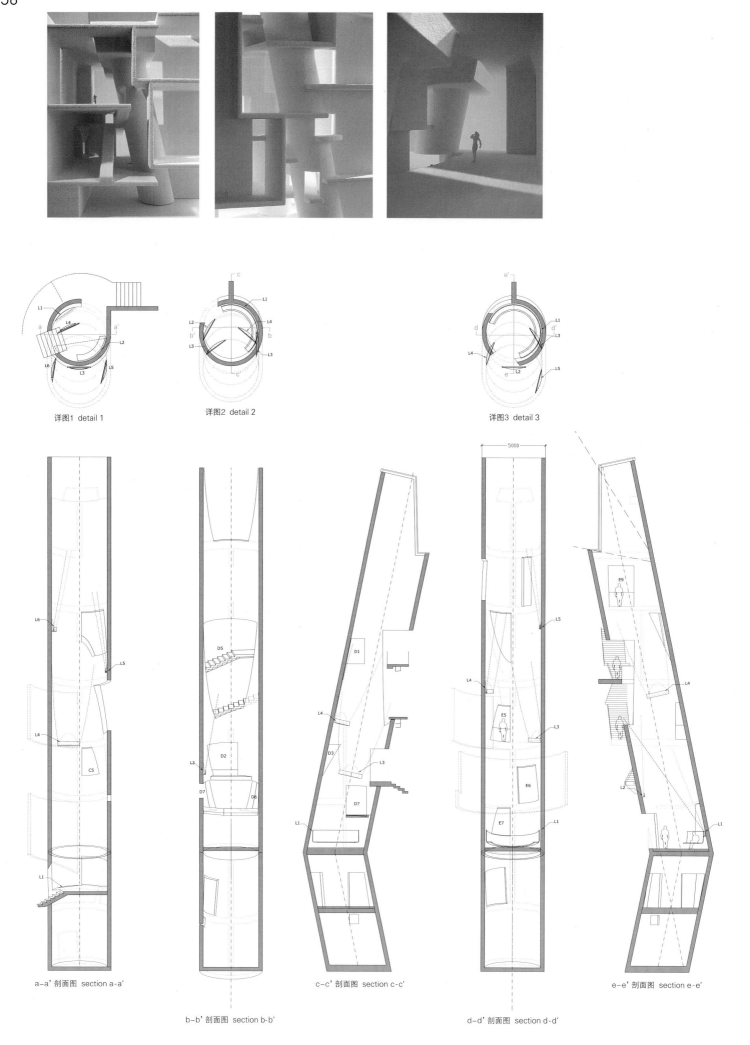

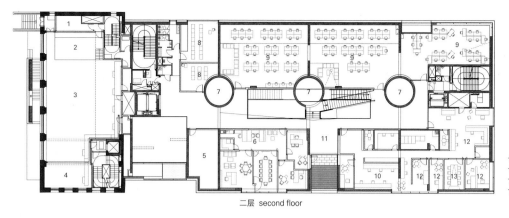

| | |
|---|---|
| 1 排练室 | 1. rehearsal room |
| 2 舞台 | 2. stage |
| 3 会场 | 3. venue |
| 4 酒吧 | 4. bar |
| 5 项目空间 | 5. project space |
| 6 董事会套间 | 6. directorate suite |
| 7 趋光空间 | 7. driven void |
| 8 总研讨室 | 8. general seminar room |
| 9 设计交流工作室 | 9. communication design studio |
| 10 设计交流员工室 | 10. communication design staff room |
| 11 非正式的自习室 | 11. informal learning space |
| 12 设计管理学院 | 12. school of design management |
| 13 会客室 | 13. meeting room |

二层 second floor

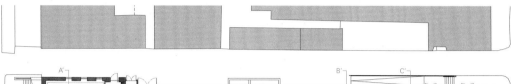

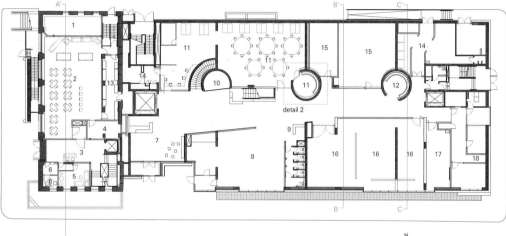

| | |
|---|---|
| 1 车间 | 1. plant |
| 2 会场 | 2. venue |
| 3 入口大厅/安保室 | 3. entry hall/security |
| 4 厨房 | 4. kitchen |
| 5 休息室 | 5. lounge |
| 6 办公室 | 6. office |
| 7 主大厅 | 7. main lobby |
| 8 麦金托什大楼窗户 | 8. window on Mackintosh |
| 9 咖啡室 | 9. cafe |
| 10 趋光空间 | 10. driven void |
| 11 数码工作间 | 11. digital workshop |
| 12 非正式的自习室 | 12. informal learning space |
| 13 酒吧/酒窖 | 13. bar store/cellar |
| 14 存储室 | 14. storage |
| 15 主研讨室 | 15. principal seminar room |
| 16 展览空间 | 16. exhibition space |
| 17 项目空间 | 17. project space |
| 18 现代纺织中心 | 18. center for advanced textiles |

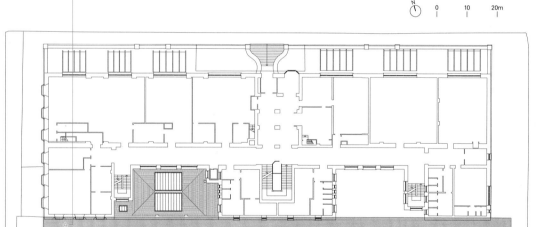

一层 first floor

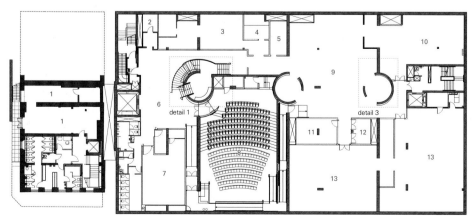

| | |
|---|---|
| 1 车间 | 1. plant |
| 2 卫生间 | 2. rest room |
| 3 现代原型设计&塑胶工作室 | 3. advanced prototyping & plastics workshop |
| 4 接待处&工具商店 | 4. reception/tool store |
| 5 电气工作间 | 5. electronics workshop |
| 6 阶梯大厅 | 6. lecture theatre lobby |
| 7 摄影工作室和暗室 | 7. photo studio and dark room |
| 8 主阶梯室 | 8. main lecture theatre |
| 9 装配空间 | 9. assembly spaces |
| 10 金属工作间 | 10. metal workshop |
| 11 技术员工室 | 11. technical staff room |
| 12 折铸亭 | 12. spray booth |
| 13 木材工作间 | 13. wood workshop |

地下一层 first floor below ground

| | |
|---|---|
| 1 纺织品工作室 | 1. textiles studio |
| 2 存储室 | 2. storage |
| 3 纺织品/缝纫和针织工作室 | 3. textiles /sewing and knitting studio |
| 4 研究生工作室 | 4. post graduate studio |
| 5 趋光空间 | 5. driven void |
| 6 编织工作间 | 6. weaving workshop |
| 7 爆光区 | 7. screen exposure |
| 8 丝网印刷区 | 8. screen printing |
| 9 染料实验室 | 9. dye lab |
| 10 员工室 | 10. staff room |
| 11 院长室 | 11. head of department |

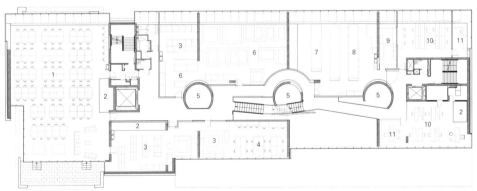

四层 fourth floor

| | |
|---|---|
| 1 温室/商店 | 1. green room/store |
| 2 会场 | 2. venue |
| 3 员工室 | 3. staff room |
| 4 画廊 | 4. gallery |
| 5 陈列室 | 5. case room |
| 6 办公室 | 6. office |
| 7 数码工作间 | 7. digital workshop |
| 8 驱光空间 | 8. driven void of light |
| 9 休息室 | 9. soft lounge |
| 10 私人餐厅 | 10. private dining |
| 11 员工餐厅 | 11. staff dining |
| 12 设计交流工作室 | 12. communication design studio |
| 13 厨房 | 13. kitchen |
| 14 调查室 | 14. survey |
| 15 主餐厅 | 15. main refectory space |

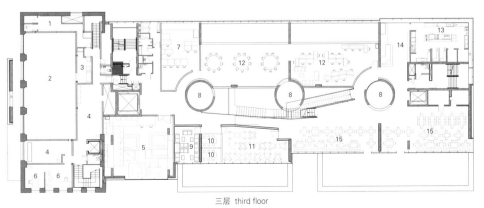

三层 third floor

# 林间小筑
# We Live in

C3往期的丛书中提及过以木材作为建筑主要材料的项目,以此来唤醒建筑技术的历史、空间类型以及国际规模经济。可能是受到生态研究的影响,如今建筑师倾向于把木材作为一种革新方法的个性元素,这较少是出于消费动机,更多是出于对地球自然平衡的考虑。

但是,本期的某些案例回归了木材,它们被转化成针对自然的新构想研究:在这个意义上,这里提出的项目似乎在寻找沉浸在景观的原因,而木材成为一种把建筑推向征服荒野的托辞。

从这个意义上来说,亨利·戴维·梭罗的作品将是描述这段荒野旅程的起点。这段旅程将谈述开放的建筑,其中木材是建成环境和自然领域共生关系的隐喻。埃托·索特萨斯和石上纯也的项目创造了一条空间实验和极端方法之间的纽带。亨利·戴维·梭罗曾企图将这一极端方法应用他的弟子中,其中包括克里斯·麦坎得勒斯,他的故事就是电影《荒野生存》的原型。

我们会在这里分析的住宅中找到此类实验。其中野外自然完全地进入到居民的生活中,这样人们很难区分是住在木质房屋中还是森林里。

Past issues of the C3 have presented projects in which wood served as both building material and protagonist, evoking a history of building technologies, types of spaces, international scale economies. Perhaps influenced by ecological research, architects today tend to prefer wood as a characteristic element of an innovative approach, less for consumerist motives and more so out of regard for the natural balance of the Earth.

This return to wood, however, is transformed, in some cases, into research aimed toward a new conception of nature: In this sense the projects presented here seem to find their immersion reason into the landscape for being, with the wood becoming a sort of pretext for pushing the buildings toward a conquest of the wilderness.

In this sense the writings of Henry D. Thoreau will be the starting point in describing this journey into the wild, a journey that will tell of an open architecture in which wood is a metaphor for this symbiosis between built and natural territory. The projects of Ettore Sottsass and Junya Ishigami create a link between spatial experimentation and the extreme approach that Thoreau sought to induce in his disciples, including Chris McCandless, whose stories are told in the film *Into the Wild*.

We will find such experimentation in the homes analyzed here, in which wild nature enters fully into the life of the inhabitants, and in which one may scarcely distinguish between living in wood, or in the woods.

Norderhov林中小屋_Cabin Norderhov/Atelier Oslo
梅里之家_Meri House/Pezo Von Ellrichshausen Architects
"狗跑"住宅_Dogtrot House/Dunn & Hillam Architects
踢马公寓_Kicking Horse Residence /
Bohlin Cywinski Jackson in association with Bohlin Grauman Miller Architects
布鲁恩别墅_Villa Bruun/Häkli Architects
环绕红桉的住宅_Karri Loop House/MORQ
林间小筑_We Live in the Wood(s)/Diego Terna

# the Wood(s)

**Hard Sun (强烈的阳光)**

"我有时梦见了一座较大的能容得下很多人的房屋，矗立在黄金时代中，材料耐用持久，屋顶上也没有华而不实的装饰，可是它只包括一个房间，一个宽阔、简朴、实用而具有原始风味的厅堂，没有天花板，也没有灰泥墙面，只有光光的橡木和桁条，支撑着头顶上较低的天堂，却足以抵御雨雪了……一个空洞的房间，你必须把火炬装在一根长竿的顶端方能看到屋顶，而在那里，有人可以住在炉边，有人可以住在窗口凹处，有人在高背长椅上，有人在大厅一端，有人在另一端，如果他们中意，有人可以和蜘蛛一起住在橡木上；这种住宅，你一打开大门就到了里边，不必再拘泥形迹……这正是在暴风雨之夜你愿意到达的一间房屋，一切应有尽有，又无管理家务之烦……同时它还是厨房、餐具室、客厅、卧室、栈房和阁楼……这座住宅，像鸟巢，内部公开而且明显；你无法前门进来，后门出去，而不看到它的房客；就是身为客人也享受了房屋中的全部自由，这并不是把你关在一个特别的小房间中，让你在里面自得其乐，使你孤零零地受到禁锢。"

亨利·戴维·梭罗，《瓦尔登湖》或《林中日月1》，1854年

## Hard Sun

*I sometimes dream of a larger and more populous house, standing in a golden age, of enduring materials, and without gingerbread work, which shall still consist of only one room, a vast, rude, substantial, primitive hall, without ceiling or plastering, with bare rafters and purlins supporting a sort of lower heaven over one's head – useful to keep off rain and snow [...] a cavernous house, wherein you must reach up a torch upon a pole to see the roof; where some may live in the fireplace, some in the recess of a window, and some on settles, some at one end of the hall, some at another, and some aloft on rafters with the spiders, if they choose; a house which you have got into when you have opened the outside door, and the ceremony is over; [...] such a shelter as you would be glad to reach in a tempestuous night, containing all the essentials of a house, and nothing for house-keeping; [...] at once kitchen, pantry, parlor, chamber, storehouse, and garret [...] A house whose inside is as open and manifest as a bird's nest, and you cannot go in at the front door and out at the back without seeing some of its inhabitants; where to be a guest is to be presented with the freedom of the house, and not to be carefully excluded from seven eighths of it, shut up in a particular cell, and told to make yourself at home there in solitary confinement.*

-Henry D. Thoreau, *Walden, or Life in the Woods 1*, 1854.

似乎只有音乐，积极向上的音乐，而非文字，可以伴随克里斯·麦坎得勒斯在阿拉斯加荒野里的废弃公交中死去的最后形象。在由西恩·潘执导的电影《荒野生存》中，事实上，Eddie Vedder演奏的吉他声（歌曲《Hard Sun》），最后陪伴着克里斯垂死的叹气，而这个美国歌手的短音暗示了这位年轻探险家的归宿。

镜头缓缓向上升起，最终向我们展示了主角的房子的延伸，不仅仅是旧公交车，还有围绕着它的区域：草地、山（作为墙体）和天空（在地平线处闭幕）。

It seems that only music, rather than words, could accompany, with a touching movement upwards, the last image of Christopher McCandless, dying in an abandoned bus in the wilds of Alaska. In the movie *Into the Wild*, directed by Sean Penn, Eddie Vedder's guitar (from the song *Hard Sun*) is, in fact, the last sound that accompanies Christ's dying sigh, whereupon only short vocalizations of the American singer indicate the end of the young explorer.

The camera slowly rises upward and finally shows us an extension of the house of the protagonist, not just the old bus, but all the territory that surrounds it: a grass floor, mountains for walls, the sky as a closing on the horizon.

It is no coincidence that McCandless was a deep admirer of Thoreau: Both sought in a genuine relationship with the landscape, the real reason for the existence of a civilized life, a life that found in a return to Nature as a new starting point for all of humanity. Both tried to escape from a dimension of civilization, in their view, that had strayed far from the true essence of humankind, diving with their own physiques into the difficulties of the wilderness, stingy of amenities, but generous in true happiness: contact with the unknown territory, still virgin, free from urban landscape modification.

麦坎得勒斯是梭罗忠实的崇拜者，这并非是巧合：两者都在寻求与景观的真正联系，一个文明生活存在的真实理由，在回归自然中找到的生活是人性的一个新的起点。

两人都想逃离文明的维度，在他们看来，那里已经远离人类真谛，推动他们自身来到充满困难的荒野，这里只有有限的设备，但是带给他们的却是真正的幸福：接触到未知的地域，这里还没有开发，不受城市景观改造的束缚。

梭罗精确地描述他以后的住所：一个简单原始的房间，没有天花板，很原始的；家中每一位成员都能够找到属于自己的地方，它可以是壁炉，窗户的凹槽，甚至是房梁；可以在暴风雨来临之际作为避难所的房子。梭罗的房子是内外部完美的融合，就地取材，不需要改造：石头就是石头，木材就是木材。

在这个等价项目（木材=木材）中，人们可以发现《瓦尔登湖》中描述的项目的独特特征：文明即自然，所以家也呈现出稍微被改造的自然的外观和特点。

如今，130年后，我们又一次在摇晃的公交中看到无限富饶与美好的生活：即使没有等同的建筑材料，沉浸在土地中的公交车已经成为土地的一部分：一个简单的房间，配有炉火、避难所、厨房、起居室、卧室，以及室外，即荒野，它也是那个小公交的延伸，且本身已成为探险者的家园。

在不接受都市生活的革命性行为中，麦坎得勒斯向我们展示，从零开始总是可能的，构成我们日常生活的日程表是可以分开的，只要生活本身被新的眼光看待，并且我们接受遭遇辽阔的大自然问题时的疲劳。

**木材，森林，家园**

麦坎得勒斯处理生活的方式比梭罗更加激进：后者不会接受这种方式，例如，拥有一个非自建的家园。从这个意义上说，梭罗进行设计，不是"为"自然，而是和它"一起"。总之，他的家，是要作为他本身的一

Thoreau describes exactly what would later be the abode of his counterpart: a single room, with no ceiling, primitive; a home where each person can find his or her own place, be it the fireplace, a recess in the window, or even the roof beams; a house that is refuge on a stormy day. The house of Thoreau is a perfect integration of outside and inside, made with building materials that are of the territory and do not need to be transformed: a stone is a stone, wood is wood.

In this equivalence (wood = wood) one can find the distinctive feature of the project described in Walden: Civilization is nature and therefore the home also takes on the appearance and features of a nature that is only slightly domesticated.

Now, 130 years later, we see again in that rickety bus the infinite richness of a good life: Even without an equivalence of building materials, the bus is so immersed in the landscape as to be an integral part of it: a single room, which serves as fireplace, shelter, kitchen, living room, bedroom and, outside, the wilderness, which acts as an extension of that little bus and itself becomes the home of the explorer.

In the revolutionary act of not accepting his urban life, McCandless shows us that it is always possible to start from scratch, that the schedule that structures our daily lives can be torn open, as long as life itself is seen with new eyes and we accept the fatigue that comes from confronting the enormity of the surrounding nature.

### Wood, Woods, Home

McCandless dealt with life in a much more radical way than Thoreau: The latter would not accept, for example, having a home that was not self-built. Thoreau proceeded with a designing attitude, working not "for" Nature, but "together" with it. His home, in short, was to be an extension of his own being, clearly in tune with the surrounding landscape.

Twenty years before the death of McCandless, an Italian architect, Ettore Sottsass, reconstructed, with surprising poverty of means but extraordinary wealth of space, the yearning for Nature that brought Thoreau to isolate himself from the world and to build his home.

In the early 1970s, Sottsass photographed a series of minimal architecture in the Spanish desert and the Pyrenees which he had built as a primordial act of appropriating the territory. One of these, entitled *Drawing A Door to Enter the Dark*, in 1973, seems to take to its extreme of Thoreau's design idea, a home that wel-

西恩·潘执导的电影《荒野生存》，2007年
Into the Wild directed by Sean Penn, 2007

个延伸，显然要与周围环境相处融洽。

在麦坎得勒斯逝世之前的二十年，一位意大利的建筑师，埃托·索特萨斯，用令人惊讶的有限方法和非凡的空间资源，再一次实现了梭罗把自己从世界中孤立并建造属于自己的住宅那样的对自然和家园的渴望。

在20世纪70年代早期，索特萨斯在西班牙的沙漠和Pyrenees山脉拍摄了一系列自建的小建筑，他建这些是出于一个原始的占有领地的目的。其中，在1973年，一座名为"画一扇门进入黑暗"的建筑，似乎达到梭罗设计理念的极限。那是一处简单但是充满热情氛围的家园，越来越大，直到其空间融入周边地区，正如麦坎得勒斯的公共汽车一样。

然而，在这里，房子的柱子是由轻量级的、带有叶子的树枝建成的。通过这种方式，索特萨斯向我们展示了一个全新的等价：木材保留其自然的外观，同时没有成为建筑材料，尽管它是定义建筑的主要元素。

在这里，木材又一次是森林和建筑。

意大利建筑师引导我们生活在一个木质的、人造的，但是自然的、小型的建筑中，这里能够延伸，简单却丰富。在这个例子中我们可以看到梭罗和麦坎得勒斯被实现的灵感：真正的生活在于与旷野的接触；一种原始的生活，再一次以创新为前提；一种生活，无论如何，作为一种艺术行为，都与麦坎得勒斯一样极端。

最终，一位日本建筑师，结束梭罗叙述的事件的回路，其中包括麦坎得勒斯和索特萨斯的经验，且从后者的极端主义中解放自己：在2008威尼斯双年展上，石上纯也展示的装置是用一系列小温室来环绕日本馆，这些温室那么小以至于几乎不存在，但却强调了种类丰富的植物和灌木的多面性存在。

就好像，多亏这种干预，建筑师才能够呈现一个植物的世界，高度人工化。在这里，自然可以找到属于自己的空间，游客可以充分享受，尽管它有城市化布局：建筑变得稀少，或者趋于消失，让位给自然，这样，

comes with simplicity but great hospitality, becoming larger and larger until its spaces integrate with the surrounding area, as did the bus of McCandless.

Here, however, the pillars of the house are lightweight tree branches with foliage; Sottsass, in this way, shows us a new equivalence: The wood retains its natural appearance and is not turned into building material, although it is the main element defining the architecture.

The wood, here, is at once forest and building.

The Italian architect leads us to live in a wood, artificial but natural, small but extended, simple but rich. In this example we see the realized aspirations of Thoreau and McCandless: a true life, in contact with the wilderness; a primordial life, based again on innovative premises; a life, anyway, conceived as an artistic act, as extreme as that of McCandless.

It is, finally, a Japanese architect who closes the loop on the events narrated by Thoreau, encompassing the experiences of McCandless and Sottsass, emancipating himself, however, from the extremism of the latter: In 2008 Venice Biennale Junya Ishigami presents an installation which surrounds the Japanese pavilion with a series of small greenhouses, so minimal as to be almost nonexistent, emphasizing the multifaceted presence of a profusion of plants and shrubs.

And it is as if, thanks to this intervention, the architect is able to present a plant world, highly artificial, in which Nature can find its own space, fully enjoyable by the visitors, despite its urban placement: The architecture becomes rarefied and tends to disappear to make room for the green, which, in this way, becomes the architecture itself.

With this installation, full of exciting poetry, Ishigami builds a programmatic manifesto of architecture able to link the words of Thoreau with the installations of Sottsass and to finally achieve the aspirations of McCandless.

With its minute architecture – pillars as light as those of Sottsass in the Spanish desert and thin, almost non-existent – and through a prevalence of natural as opposed to built elements, the installation seems to live in a place of pure symbiosis between interior and exterior. Here there is no distinction between the life of the person and of nature, between domestic and urban, between green and built.

Perhaps it is not a purely wild territory, but even with light touches of "domestication" the Venice exhibition seems to revive a feeling of inclusion between architecture and landscape in which the ar-

布鲁恩别墅，库奥皮奥，芬兰
Villa Bruun in Kuopio, Finland

照片提供：©Häkli Architects/Jussi Tiainen

自然就成为建筑本身。

这个装置充满了诗意，石上纯也构建了纲领性的建筑宣言。这一宣言能够将梭罗的语言与索特萨斯的装置结合，最终实现麦坎得勒斯的雄心壮志。

在这座微小的建筑里——柱子就像索特萨斯在西班牙沙漠里构建的建筑的柱子那么轻和薄，几乎不存在。通过与建筑元素相反的，即自然元素的流行，这个装置似乎处在一处内外共生的地方。在这里，没有个人生命和大自然生命的区别，没有本地和城市生活的区别，也没有自然和建筑的区别。

也许它不是一处纯粹的野生领土，但即使有轻微的"改造"，威尼斯的展览似乎也能在建筑与景观之间纳入感情，建筑本身成为一个回归自然、自然成为建筑的隐喻。

### 木质房屋

在这里介绍的项目中，主要的建筑材料木材，成为代表自然生活的隐喻：在一种自然的条件下，因为木材的自然属性，借助木材模拟丛林中的生活，也算是一种权宜之计。当它与石上纯也的温室的突出特点相近时，人们便可以观察到试图描述索特萨斯设计演变中的一些常见的元素：住宅空间的特点是企图从市区逃离，成为一处在自然的环境中有点复杂的避难所。

索特萨斯和石上纯也的项目追求的是野性与驯服之间的富有成果的对话，内外相互渗透被放大到最大：正如意大利建筑师的装置，木材有其隐喻的重要性。本章所展现的项目可以在三个宏观类里进行观察，它们基本上反映一棵树的成长，从树根到树冠，摇身变成树林，与此同时，变成住宅。

### 树根

MORQ的环绕红桉的住宅和Häkli建筑师事务所的布鲁恩别墅都"降落"在地面上来寻找联系，环绕自然，仿佛直接从土壤中长出。环绕红桉的住宅潜入树木的空隙中，成为一个有生命的生物体。它的形状不是由

chitecture itself becomes a metaphor for a return to nature, which itself becomes the architecture.

### A House of Wood

In the projects presented here, wood, the main building material, becomes a metaphor for a natural life: The wood is an expedient by which is proposed a life in the woods, in a condition, therefore, of naturalness. It is possible to observe some common elements that tend to describe the evolution embodied in the designs of Sottsass as it approaches the features highlighted in the greenhouses of Ishigami: residential spaces, the peculiarity of which is their attempt to escape from the urban and to become a shelter, more or less complex, in a strongly natural environment.

What is sought is a fruitful dialogue between the wild and the tame, in the manner of the projects of Sottsass and Ishigami, in which interpenetration between inside and outside is at a maximum: As in the installations of the Italian architect, then, the material of wood assumes an importance that is also metaphorical. The presented project can be observed under three macro categories, which essentially reflect the growing of a tree, from the base to the top, transforming itself into woods and, simultaneously, into homes.

### Roots

The Karri Loop House by MORQ and the Villa Bruun by Häkli Architects alight on the ground and search for contact, wrapping around nature as if arising directly from the soil. The Karri Loop House insinuates itself into the spaces left by the trees, becoming a kind of living organism whose shape is defined not by an abstract geometry, but by the very presence of the territory, as if the accidents of nature had directly planned the dwelling's morphology. The Villa Bruun opens onto its surroundings via a large patio, where the cover acts as a filter between landscape and home; here unfolds a hybrid "stay", a domesticated nature, which highlights the fragility of the limits: What is the boundary between inside and out? The project is built precisely around this ambiguity; if in the first house the formal accident was defined by the trees, here the geometric rigidity is continually challenged by the difficulty of defining where one is: Are we in the house? Are we out? Are we in the woods?

踢马公寓，不列颠哥伦比亚，加拿大
Kicking Horse Residence in British Columbia, Canada

Norderhov林间小屋，Norderhov，挪威
Cabin Norderhov in Norderhov, Norway

抽象的几何定义，而是由地域所定义，仿佛大自然的意外曾直接策划该住宅的形态。布鲁恩别墅则通过大型庭院使建筑向四周开放，在那里，覆顶成为景观和家园之间的过滤器。这里展现了一种复杂的"逗留"，被改造的自然，突出了界线的脆弱性：内部和外部之间的边界是什么？项目精准地围绕这一模糊性进行建造；如果在第一座住宅中，意外事件是由树定义的，那么这个项目的几何外形的刻板性不断地被定义一个人的位置所带来的困难性质疑：我们是在房子里吗？我们是在外面吗？我们是在树林中吗？

### 叶子

由Dunn&Hillam建筑师事务所设计的"狗跑"住宅，Bohlin Cywinski Jackson和Bohlin Grauman Miller建筑师事务所联合设计的踢马公寓明确重申这类住房的概念，它们各自以自己的方式，来突出房屋的覆盖面，作为住宅本身的一个重要因素。

在这两个项目中，屋顶就像一棵树的树冠，成为识别建筑的一个因素。它暗含了建筑的本质，并且建立了房子空间与外面自然空间之间的对话。房子的屋顶承担了不仅作为覆盖功能的重要性：它们就像梭罗说的那样，是一个低矮的天堂，可以遮盖但不会隐藏外部的视野。

### 林间小筑

Pezo Von Ellrichshausen建筑师事务所设计的梅里之家和Oslo工作室设计的Norderhov林中小屋找到了麦坎得勒斯急切搜索的荒野小径的尽头，它们受到了梭罗例子的影响：木材作为原料，成为这个项目特有的规范，一个连续的表面，似乎直接来自于周围的树林。它似乎实现物质、材料和隐喻的融合：如果在石上纯也的展览中，其作品是通过寻找极端的亮度使建筑融合在绿色里来实现的，那么在这里，就是大量的木材成为主角，充满意义，伴随着不寻常的厚重感，它装饰着内部（在挪威的项目中具有流动性；在智利的项目中具有坚硬性），并将树林、自然拉入建筑的内部，迫使它们与人类共存，且获得些许诗意的空间，现在看来，这些空间更具有野性。

## Fronds

The Dogtrot House by Dunn&Hillam Architects, and the Kicking Horse Residence by Bohlin Cywinski Jackson in association with Bohlin Grauman Miller Architects clearly reaffirm the concept of this sort of shelter, highlighting, each in its own way, the coverage of the houses as a significant element of the dwelling itself.

In all two projects the roof, like the crown of a tree, becomes an element of recognition for the building. It connotes the essence, establishes a dialogue between the spaces of the house and those of the nature outside. The roofs of the houses, then, assume an importance that transcends the function of mere cover: Somehow they return to the words of Thoreau in which he describes a lower heaven that can cover but does not close off a vision of the external.

## Finally, we live in the wood(s)

The Meri House by Pezo Von Ellrichshausen Architects and the Cabin Norderhov by Atelier Oslo find the end of the path taken by McCandless in his anxious search into the wild, influenced by the example of Thoreau: Wood as material becomes here the characterizing code of the project, a continuous surface that seems to derive directly from the woods surrounding the house. It seems to realize a merger between matter, material and metaphor: If in the Ishigami exhibition the result was obtained through a search for extreme lightness that softened the architecture in green, here it is the mass of wood, full of meaning, that is the protagonist. With unusual heaviness, it upholsters the interiors (fluidly, in the Norwegian project; rigidly, in the Chilean one) and drags the woods, nature, inside the building, forcing a co-existence with the human, but achieving a light poetry in the space, which now seems, if even just a little, more wild. Diego Terna

这个项目位于Krokskogen森林, Hønefoss镇的外面。它坐落于一个陡峭的斜坡上, 这里能将Steinsfjorden湖的美景尽收眼底。由于场地的风较大, 小屋围绕着几个室外空间设计, 这些空间能提供避风场所, 且小屋可以在一天中的不同时间被太阳照射到。

内部空间是个连续的整体。4mm厚的桦木胶合板将天花板和曲形墙面覆盖成连续的表面。地面顺地势延伸, 并将平面划分成几个层次, 来定义不同的功能区。楼层之间设有台阶, 提供了各式各样的坐卧区。

房子中心的壁炉位于主入口层地面, 给人以在景观中生起篝火的感觉。从房子的各个楼层都能够看到壁炉, 你可以远观, 也可以在它旁边躺下。

大型玻璃幕墙位于客厅和用餐区。玻璃的框架很细致地避免从房间内部看到, 这样做就可以与外部自然建立更加直接的联系。

在屋外, 矩形小屋的墙面和屋顶覆盖着20mm厚的玄武岩石板, 其形式类似挪威传统的木覆层。

这座林中小屋主要是由预制构件构成的, 主结构为层压木板, 底层为Kerto胶合板。Kerto板由数控碾磨, 定义了小屋内外的几何形状。小屋由固定在岩石中的钢筋支撑, 同时为稳定起见, 在壁炉的下面建了一个小型混凝土地基。

# Norderhov林中小屋
Atelier Oslo

## Cabin Norderhov

The project is located in Krokskogen forests, outside the town of Hønefoss. Its location on a steep slope gives a fantastic view over the Lake Steinsfjorden. The site is often exposed to strong winds, so the cabin is organized around several outdoors spaces that provide shelter from the wind and receives the sun at different times of the day.

The interior is shaped as a continuous space. The curved walls and ceilings form continuous surfaces clad with 4mm birch plywood. The floor follows the terrain and divides the plan into several levels that also define the different functional zones of the cabin. The transitions between these levels create steps that provide various places for sitting and lying down.

The fireplace is located at the center of the cabin, set down in the floor of the main access level. This provides the feeling of a campfire in the landscape. Seen from all levels in the cabin, you can enjoy the fireplace from far away or lie down next to it.

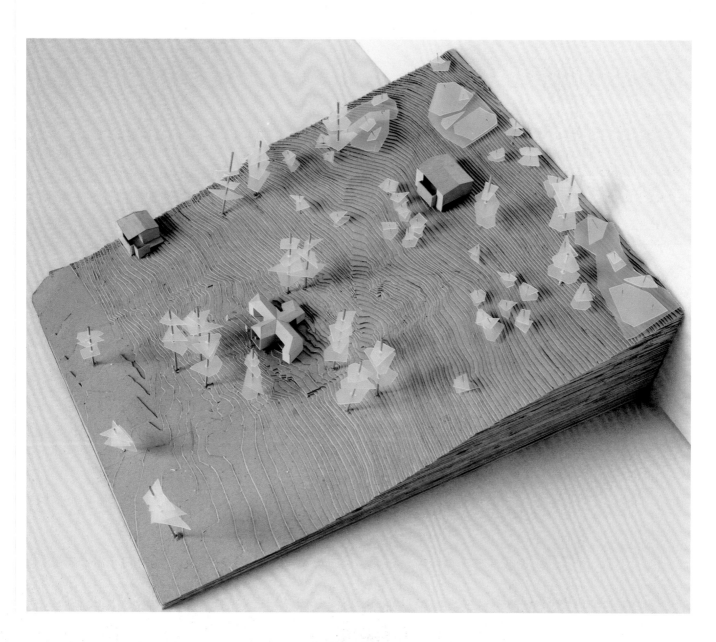

Large glass walls are located in the living and dining areas. The frames of the glass are detailed carefully to avoid seeing it from the inside. This creates a more direct relationship with the nature outside.

Outside, the cottage has a more rectangular geometry and the walls and roofs are covered with 20mm basalt stone slabs laid in a pattern similar the ones often used for traditional wooden claddings in Norway.

The lodge consists mainly of prefabricated elements. The main structure is laminated timber completed with a substructure of Kerto construction plywood. The Kerto boards are CNC milled and define the geometry both externally and internally. The cabin is supported by steel rods drilled directly into the rock, supplemented with a small concrete foundation under the fireplace for stabilization.

项目名称：Cabin Norderhov / 地点：Norderhov, Norway
建筑师：Atelier Oslo
项目团队：Nils Ole Bae Brandtzæg, Thomas Liu, Marius Mowe, Jonas Norsted, Juan Ruiz, Bosheng Gan, Sveinn Thorarinsson, Emmanuel Ferm
顾问：Sivilingeniør Ole Morten Braathen, Kåre Wærnes, Rolf Evensen, Fredrik Eng
主承包商：Byggmester Bård Bredesen
用地面积：1,500m² / 总建筑面积：55m² / 有效楼层面积：80m²
设计时间：2012 / 施工时间：2013 / 竣工时间：2014
摄影师：Courtesy of the architect - p.68, p.69
©Lars Petter Pettersen(courtesy of the architect) - p.67, p.70, p.71, p.74

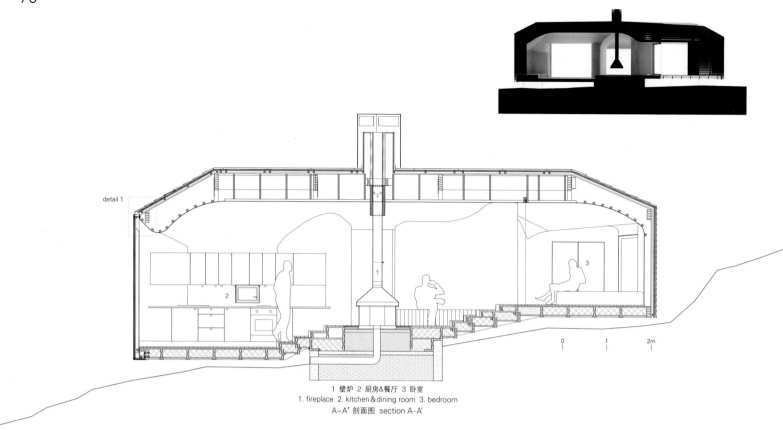

1 壁炉  2 厨房&餐厅  3 卧室
1. fireplace  2. kitchen & dining room  3. bedroom
A-A' 剖面图  section A-A'

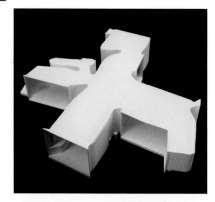

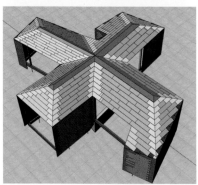

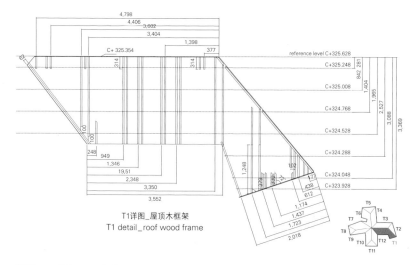

T1详图_屋顶木框架
T1 detail_roof wood frame

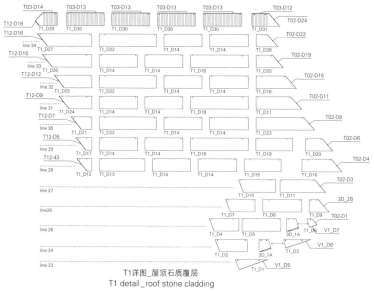

T1详图_屋顶石质覆层
T1 detail_roof stone cladding

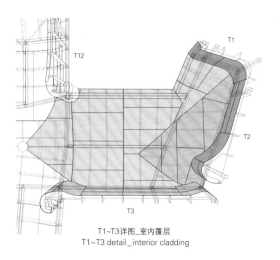

T1~T3详图_室内覆层
T1~T3 detail_interior cladding

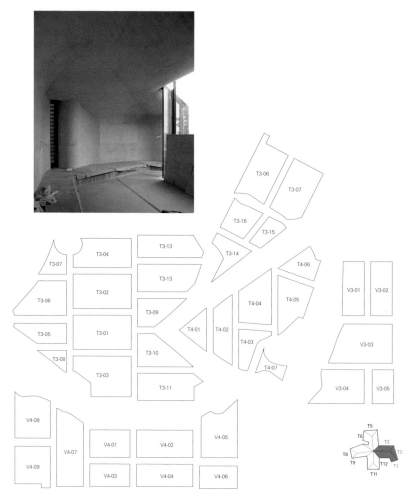

T1~T3详图_室内覆层图形　T1~T3 detail_interior cladding figure

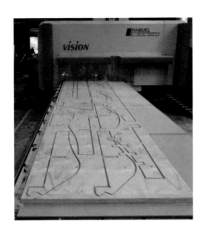
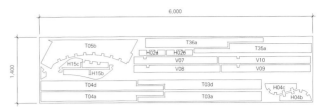

Kerto板类型1 Kerto plate type 1

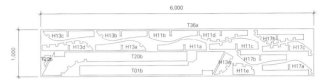

Kerto板类型2 Kerto plate type 2

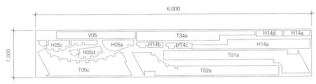

Kerto板类型3 Kerto plate type 3

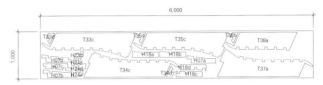

Kerto板类型4 Kerto plate type 4

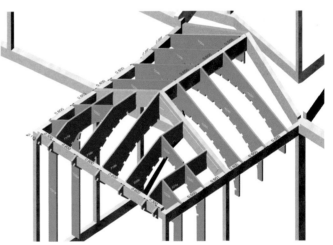

T11结构_屋顶 T11 structure_roof

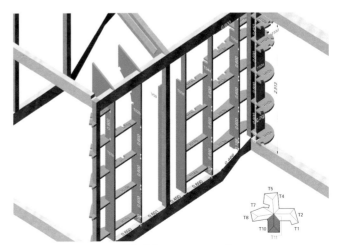

T11结构_墙体 T11 structure_wall

a-a' 剖面图 section a-a'

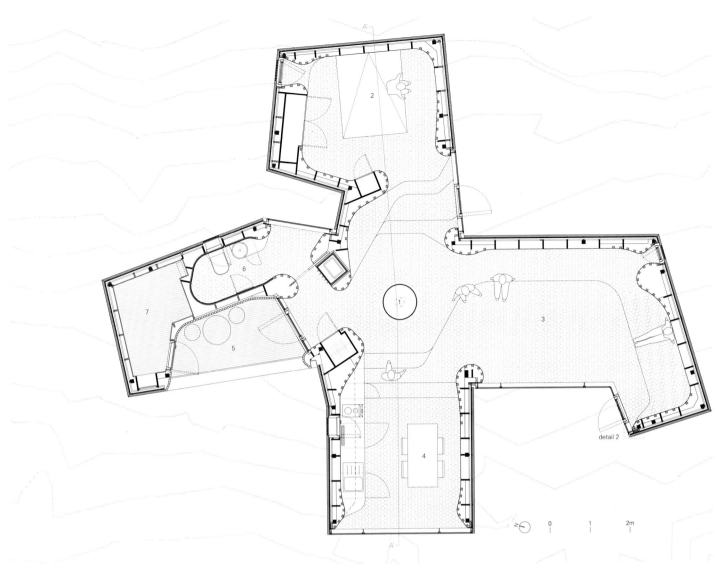

1 壁炉 2 卧室 3 起居室 4 厨房/餐厅 5 门廊 6 浴室 7 存储室
1. fireplace 2. bedroom 3. living room 4. kitchen/dining room 5. porch 6. bathroom 7. storage
一层 first floor

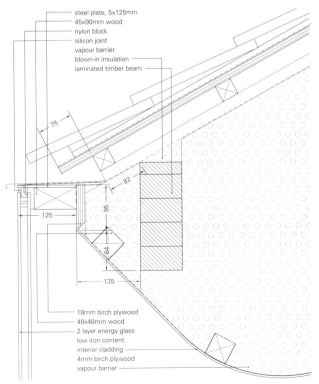

详图1 detail 1

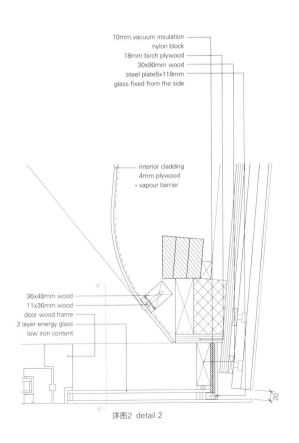

详图2 detail 2

# 梅里之家
Pezo Von Ellrichshausen Architects

最初建筑师计划这栋房子要永久使用,但是在家庭发生一些重要变化后,它被改造成这栋别样的住宅——不再是钢筋混凝土结构,而是减少了三分之一的体量并作临时使用。建筑师不只是尽量保持其光鲜自在的样子,而是在原有建筑的占地面积内真实地勾勒它,而非沿袭以往的项目。在现有的相同人工矮墙的基础上,这栋新房子沿水平方向扩大面积,以遵循山脚下蜿蜒的河流。这座建筑的主体位置通过相同房间的有规律的顺序,在其内部建立了一个双排铰接面。开阔的景观的对面是一处连续的公共功能区,而靠近山的一面是另外一处非连续的个性化功能区。每一面都有5个房间,外墙的洞口系统与方位和功能毫无关联,而另外一个内墙上的孔径体系是与房间之间的隐私程度相关的。此外,住宅还有一个天花板孔径系统,呈对角线状,给人增加平面进深的视觉体验。它使前文提到的两个系统形成一个整体。整栋房子的中心被每侧统一房间的一对正切齿孔所模糊。穿孔的天花板延伸出室内,并通过阴影分隔了房间的四面。该结构整体由浸渍松构成。外墙由染色的、未加工的松板覆盖,而内墙由柔和、平坦的木板覆盖,墙上挂着继承下来的Petoruti的绘画作品。Musil说任何想接近其他物体的事物都像是由橡皮筋拉着,在延展的时候绷紧。否则,最后它们可能相互交叉。似乎在任何活动中或者任何动作中,都幸运地有个从不让我们为所欲为的橡皮筋。

## Meri House

Even though the house the architects originally projected was for a permanent use, after important changes in the family, it was transformed into this other house, no longer in reinforced concrete, a third smaller and for a temporary use. The architects did not only try to keep its luminous and relaxed presence but to trace it literally on top of a kind of footprint of the previous project. On top of the same existing artificial podium, the new house extends itself horizontally as an attempt to follow the flow of a sinuous river at the foot of the hill. The dominant position of this piece

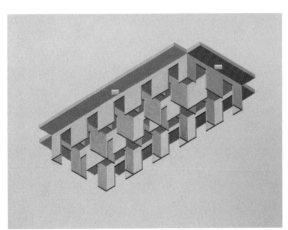

项目名称:Meri House
地点:3rd bridge, Florida, Chile
建筑师:Mauricio Pezo, Sofia von Ellrichshausen
合作者:Diego Perez, Valentina Chandia, Simon Guery, Philpe Kempfer, Orlando Hartmann, Luisa Rocco
建造商:Ricardo Ballesta
结构顾问:Luis Mendieta / 建筑服务:Luis Valenzuela, Daniel Garrido
甲方:Maria Rosa Gonzales, Alejandro Woloszyn
用地面积:180,000m² / 总建筑面积:170m²
结构:Impregnated pine wood
室外饰面:painted rough pine boards
室内饰面:untreated pine, eucalyptus boards
造价:USD 560/m²
设计时间:2013 / 施工时间:2013—2014
摄影师:courtesy of the architect

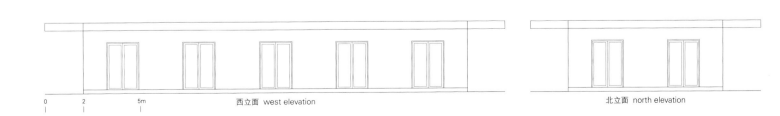

西立面 west elevation　　北立面 north elevation

establishes a double front articulated in its interior by a regular sequence of identical rooms. Towards the open landscape there is a continuous arrangement of common functions and to the immediate hill there is another discontinuous one with individual functions. Each side has five rooms with a system of openings in the exterior walls indifferent to orientation or program and another system of apertures in the interior walls that is relative to the degree of privacy from one room to the other. Both sequences are structured by another system of ceiling apertures that forces diagonally the perception of the plan's depth. The center of the whole house is blurred by a couple of tangential perforations that unify the rooms at every side. The perforated ceiling extends beyond the interior and separates each of the four sides with heavy shades. The totality of the structure is impregnated pine. The exterior walls are clad with dyed rough pine boards and the interiors with soften and flat boards, with the Petoruti painting inherited by the family. Musil said that every thing that wants to closely approximate another has an elastic band tied to it, which tenses when extended. If not, things could end up intersecting each other. It seems that in every movement, in every activity, there is an elastic that fortunately never lets us entirely do what we want.

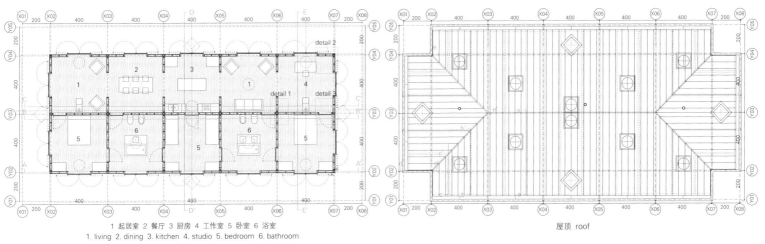

1 起居室 2 餐厅 3 厨房 4 工作室 5 卧室 6 浴室
1. living 2. dining 3. kitchen 4. studio 5. bedroom 6. bathroom
一层 first floor

屋顶 roof

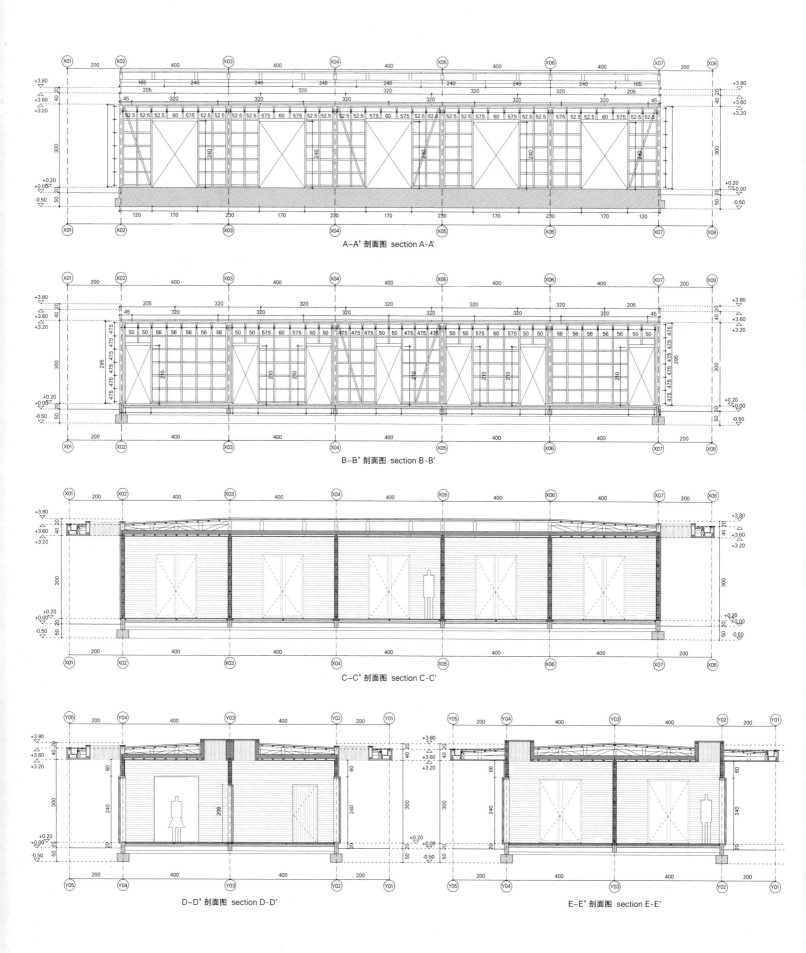

A-A' 剖面图 section A-A'

B-B' 剖面图 section B-B'

C-C' 剖面图 section C-C'

D-D' 剖面图 section D-D'

E-E' 剖面图 section E-E'

a-a' 剖面图_典型桁架N1
section a-a'_typical truss N1

b-b' 剖面图_典型桁架N2
section b-b'_typical truss N2

c-c' 剖面图_典型桁架N3
section c-c'_typical truss N3

d-d' 剖面图_典型桁架N4
section d-d'_typical truss N4

e-e' 剖面图_典型桁架N5
section e-e'_typical truss N5

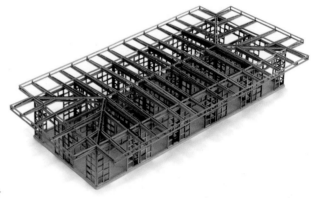

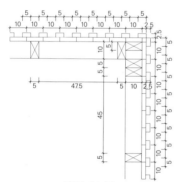

详图2 detail 2

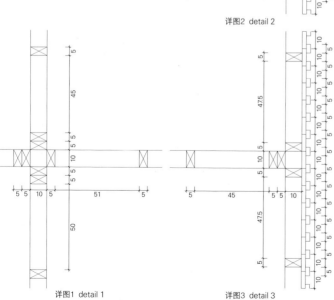

详图1 detail 1          详图3 detail 3

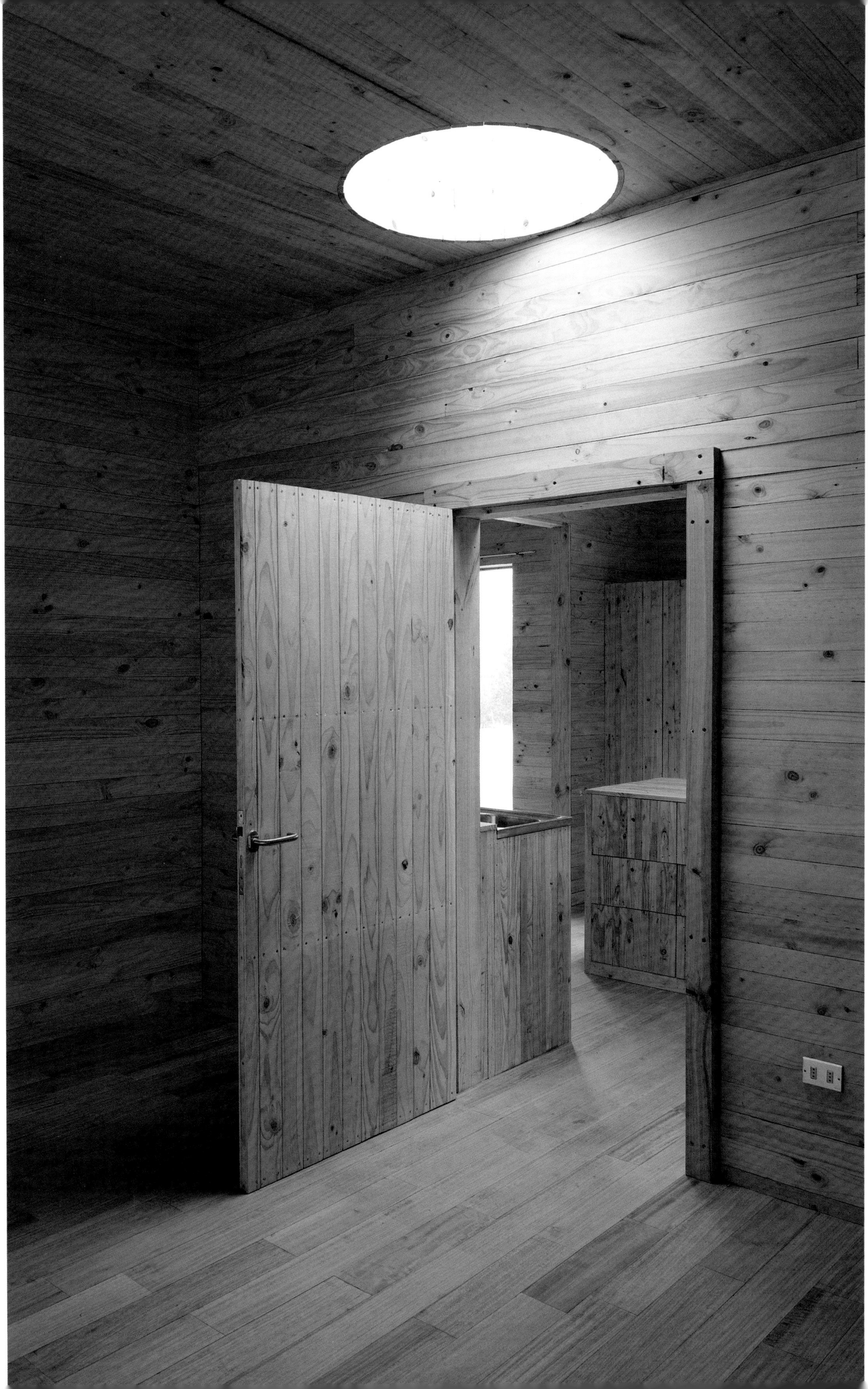

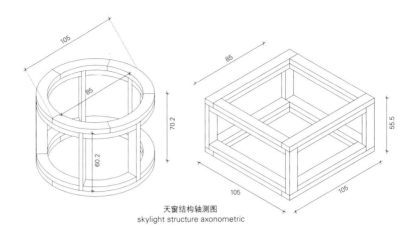

天窗结构轴测图
skylight structure axonometric

# "狗跑"住宅
Dunn & Hillam Architects

住宅建造——融入历史肌理　Dwell How – We Live in the Wood(s)

　"狗跑"住宅是一个永久的露营地，它的建筑形式可以追溯到早期农民或者渔民建造的一间式小型房子。随着家庭成员的增加，业主决定在原有房子的旁边再建造另一座小型住宅，先后两座房子共用一个屋顶。在房屋设计的演化中，建筑师重读了《杀死一只知更鸟》这本书，它为美国南部的"狗跑"住宅提供了参考。"狗跑"的名字起源于当"老狗太热而跑不动"时，有屋顶的通道，可以提供完美的庇荫，让狗或者人类远离烈日的炙烤。令建筑师欣慰的是，他们成功地建造了这样一座房子。

　　该项目应甲方的需要，是对传统的"狗跑"住宅的升级。甲方是喜欢长期徒步旅行和露营的一家人，他们要求建筑师为其设计一座度假房屋，建在他们最喜欢的营地之一延伸出来的道路上（位于新南威尔士州的南部海岸）。甲方的想法是："我们十分热爱露营，但却不希望在每个假期末尾都要打包行李，我们要的是一处永久性的文明露营地。"根据他们的想法，最终的设计方案是，建筑对周围环境影响较低，将分散的三个房间紧凑地安排在厨房和篝火周围，屋顶庇护了下面的房间，其间设置一间开放的浴室。将两座住宅分开的是一条宽敞的、开放式的遮蔽走廊。

　　住宅内房间虽然是分开的，但它们相互之间、与周围景观之间却有着紧密的联系，好像一家人在露营地上支起的一个个帐篷。中间的空地用来聚会、餐饮、读书，客户可以在徐徐微风中享受周围开阔的美景。如果想从一个房间走到另一个房间，就必须穿过室外，这样，他们总是能看到周围的景观，并了解外面的天气。建筑师很难有机会建造具有澳大利亚本土营地风格的、早期居民传统房屋结构那种周到而深刻理解所处位置的特点的房屋。而该项目就为设计师提供了这样一个机会，细节设计及材料的选择也因此应运而生。总的来说，建筑师的任务就是对项目所处位置、甲方的需要、预算等做出适当的回应，所有这些都要做得适度、简单，却不能太直截了当。

　　"狗跑"住宅，一切为你的需求而存在。不张扬，不伪装，富有诗意。

## Dogtrot House

The Dogtrot House is a permanent campsite. The form of this building can be traced back to the early one-room cabins that were built by farmers and fishermen. As the family grew another cabin would be built and connected with a common roof. During the evolution of the design for the house the architect was re-reading *To Kill A Mockingbird*, which made reference to the vernacular "dog trot" houses of the South American. The name comes from when the "old dog was too hot to trot", the covered breezeway provided the perfect escape from the heat of the day. The architects realised with some joy that this is what they had. The house represents an upgrade for the clients, who are a family

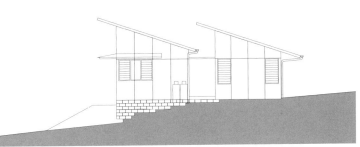

西立面 west elevation

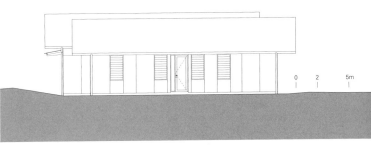

南立面 south elevation

项目名称：Dogtrot House / 地点：New South Wales, Australia
建筑师：Dunn & Hillam Architects
设计团队：Ashley Dunn, Lee Hillam, Linden Thorley, Jonathan Temple
结构工程师：John Carrick
建造商：Smith and Primmer Builders, Moruya NSW
用地面积：1,019m² / 总建筑面积：133m² / 有效楼层面积：104m²
设计时间：2012 / 施工时间：2013.4—2013.12
摄影师：©Kilian O'Sullivan(courtesy of the architect)

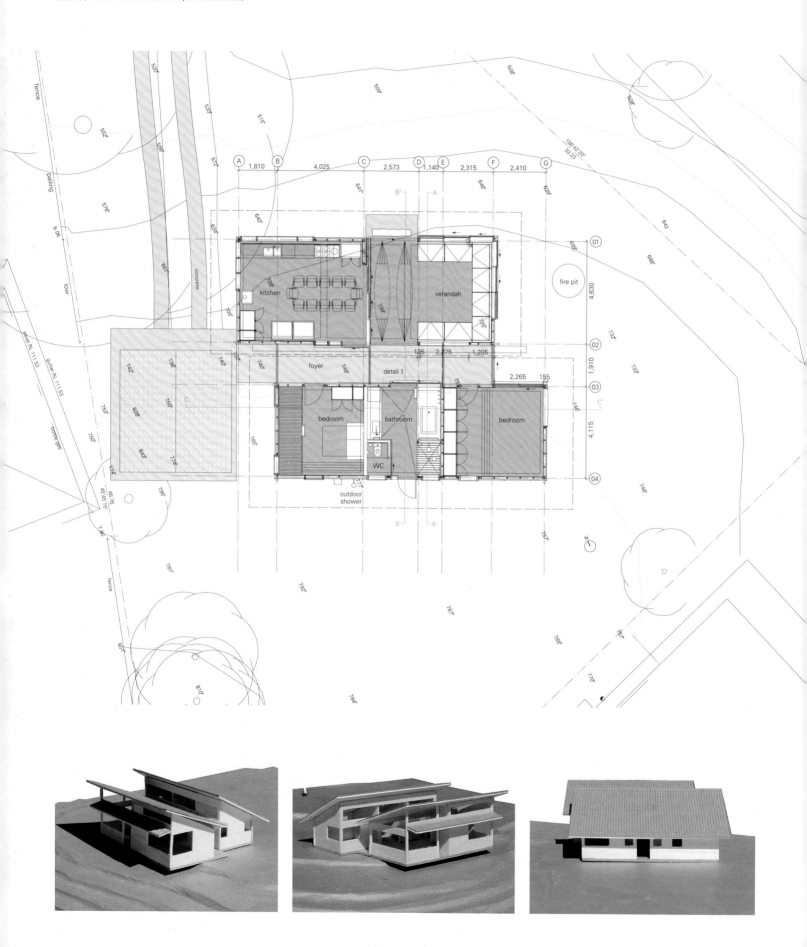

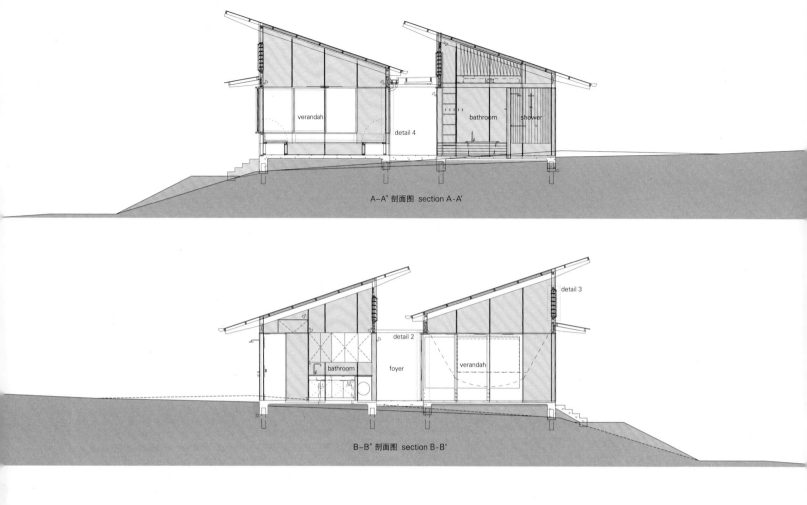

A-A' 剖面图 section A-A'

B-B' 剖面图 section B-B'

C-C' 剖面图 section C-C'

of long term and committed campers. They asked the architects to design a holiday house up the road from one of their favourite campsites on the south coast of NSW. Their brief was, everything they loved about camping, without the pack-up at the end of every holiday; a permanent, civilised campsite. The resulting building is a low impact, compact arrangement of three discrete rooms around a kitchen and campfire. The roof shelters the pavilions with an open bathroom between. Separating the pavilions is the wide, open-ended "dogtrot" corridor.

The rooms are separate but in a strong relationship to one another and the landscape, like a family of tents at a campsite. The space made in the centre is for gathering, eating, reading and is open to the landscape and the breeze. One has no choice but to be outside when moving between rooms, in this way one is always made aware of the landscape and the weather. Architects are rarely presented with an opportunity to site their buildings with the delicate and deep understanding of place that is evident in the Indigenous Australian camp and subsequently the vernacular structures of the early settlers. This project gave the architects that opportunity and the detailed design and material choices were all informed by this. Overall the architects were concerned with making an appropriate response to the place, the clients brief and the budget, all of which were modest, simple but not straightforward.

The Dogtrot House is a house that is everything you need and nothing you don't. It is humble, poetic and without pretence.

detail 1

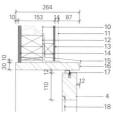

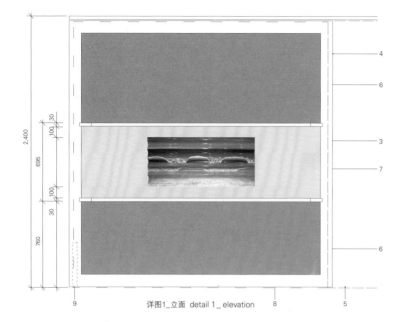

详图1_立面 detail 1_ elevation

a-a' 剖面详图 section detail a-a'

1. hardwood transom rails
2. artwork
3. FC sheet
4. hardwood door sash
5. dotted line indicates location of door when open on running gear
6. bronze mesh
7. artwork 980mmx435mmx60mm
8. dotted line indicates clear opening of doorway inside frame
9. lockable center flushbolt to exterior face
10. 90mm stud frame
11. 14mmx30mm hardwood cover strip to cladding joint
12. sarking
13. 6mm fibre cement sheet cladding
14. 12mm jointed lining boards
15. folded zincalume flashing
16. continuous hardwood door head
17. continuous stainless steel angle head guide running in saw cut in top of door, saw cut to be enlarged through middle, part of sash to reduce chances of door binding
18. bronze mesh to bushfire requirements
19. 90x30mm hardwood transom rail with rebate to accept FC sheet
20. 2x sheets of 6mm clear finished FC sheet bonded together. screw fix to transom rails and cover fixings with 30mmx10mm hardwood cover strip
21. enamel paint and fibreglass on curved corrugated steel
22. nylon filament brush seal
23. brio timberoll 200 bottom rolling sliding gear running in 9mm radius channel in bottom of door sash with 914b bottom rail

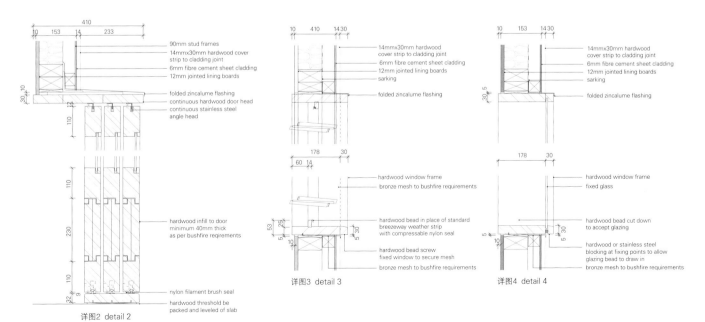

详图2 detail 2

详图3 detail 3

详图4 detail 4

## 踢马公寓

Bohlin Cywinski Jackson
in association with Bohlin Grauman Miller Architects

在加拿大落基山脉的踢马滑雪度假基地，有一处323m²的聚集地，这是一处能容纳五口之家聚会的场所。甲方期望建造一处可以亲近自然，进行季节性娱乐，灵活并能容纳更多家人和朋友，可供多达14人过夜的空间。

该公寓所处的斜坡毗邻滑雪道，四周被高山和满是云杉、白杨的森林包围着。建筑包含两部分：房子北部排列着的一些密集的小房间，用做卧室和洗浴间；一处开敞空间，作为客厅和餐厅，使人在室内便可以欣赏到外面的壮丽山景。玻璃墙体连接着这些建筑形式，公寓的主入口位于低层，而高层有一个平台可以通向西边的滑雪道。住宅的功能规划周密，给人一种开放和通透之感，同时阻挡了邻居望向这边的视野。

建筑内的休息区处在悬臂上的一处狭长空间内，它的下方是建筑物的底层，其裸露的混凝土柱用木板装饰，内设车库、前厅和娱乐室。外露的木质结构和高高的悬挑屋顶充分表现出了山中小屋精湛的构造品质。主入口位于两堵混凝土墙体之间，钢材和红木制成的楼梯也可通向楼上的生活区。一系列钢柱呈线状，沿着主门厅进行延伸，突出了卧室屋顶的角度，屋顶在顶尖处又折叠下来，形成一面带有可控通风口的铰接式金属墙，这样，阳光和新鲜空气都能进入阁楼空间。阁楼内摆放着双层床，也可以让孩子或客人灵活地安排睡眠。

室内，高大、板式混凝土壁炉对悬于树林地面层一定高度的客厅和餐厅起到了固定的作用，这样，整座建筑的排水系统就可以畅通无阻。光滑的花旗松木胶合板覆盖的墙壁和天花板延伸到外部，框出了一道高山风景，且对户外甲板起到保护的作用。外层覆有染黑的纯色雪松壁板，与外部风景形成鲜明对比。颜色鲜亮的纤维水泥板也为立面的天然木饰面增光添彩。

这座家庭寓所利用强有力的形式和唤起共鸣的材料来拥抱着大自然。

### Kicking Horse Residence

Situated at the base of the Kicking Horse Ski Resort in the Canadian Rocky Mountains, this 3,500 sq. ft. residence is a gathering place for an active family of five. The clients desired a space that would provide a direct connection to the landscape for seasonal recreation and the flexibility to accommodate larger groups of family and friends, with beds for up to 14 people.

The sloping site is adjacent to a ski trail and surrounded by an alpine forest of spruce and aspen. The house is arranged as two elements on the site: a dense bar along the northern edge containing the sleeping and bath spaces, and an open shell with living and dining spaces oriented toward the extraordinary mountain views. A glass volume links these forms, with the main entrance at the lower level and an upper landing for ski access on the west

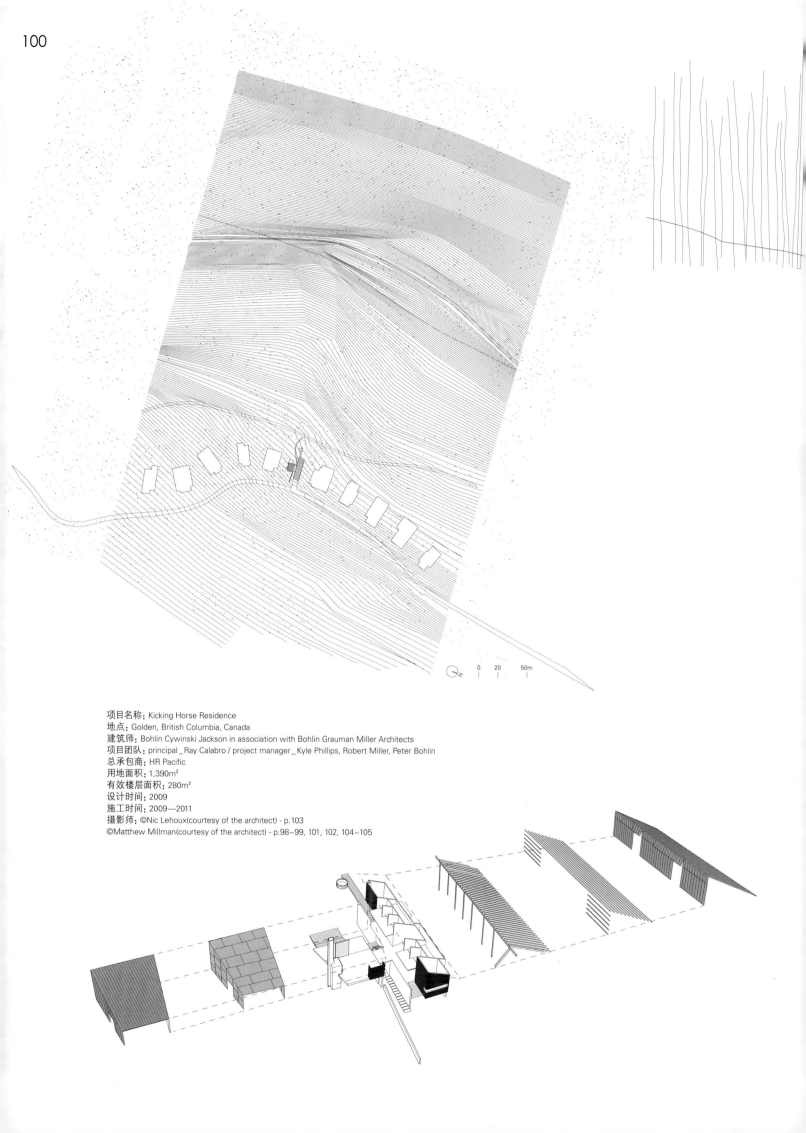

项目名称：Kicking Horse Residence
地点：Golden, British Columbia, Canada
建筑师：Bohlin Cywinski Jackson in association with Bohlin Grauman Miller Architects
项目团队：principal _ Ray Calabro / project manager _ Kyle Phillips, Robert Miller, Peter Bohlin
总承包商：HR Pacific
用地面积：1,390m²
有效楼层面积：280m²
设计时间：2009
施工时间：2009—2011
摄影师：©Nic Lehoux(courtesy of the architect) - p.103
©Matthew Millman(courtesy of the architect) - p.98~99, 101, 102, 104~105

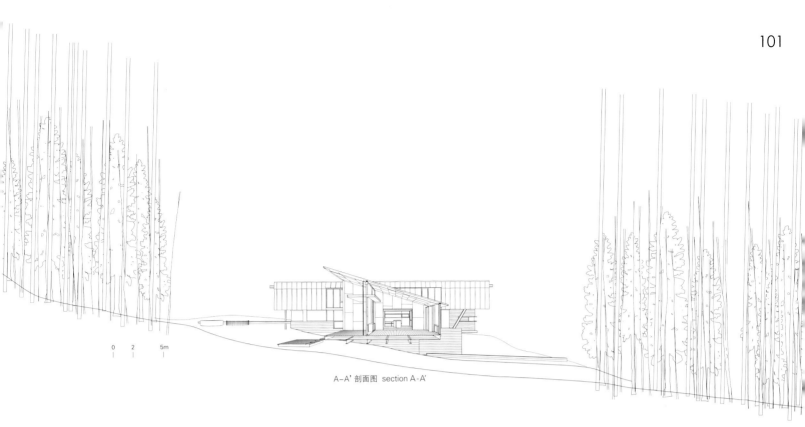

A-A' 剖面图 section A-A'

side. Careful positioning of program enables a sense of openness and transparency while screening the neighboring homes from view.

The linear form of the sleeping spaces cantilevers over a board-formed concrete base containing the garage, mudroom, and playroom. The exposed wood structure and deep roof overhangs express the tectonic qualities of a mountain cabin. The primary entry is located between two concrete walls, with a mahogany and steel stair leading to the living spaces above. A single line of steel columns extends along the main hallway, emphasizing the angle of the bedroom roof, which folds over the peak to become an articulated metal wall with operable vents, bringing light and air into the loft spaces. The lofts contain bunk beds that allow flexible sleeping arrangements for children or guests.

Anchored by a tall, board-formed concrete fireplace, the geometric form containing the living and dining spaces floats above the forest floor, allowing natural drainage to flow uninterrupted through the site. Smooth Douglas fir plywood-clad wall and ceiling planes extend to the exterior, framing alpine views and sheltering an outdoor deck. The exterior is clad in a combination of black-stained and clear cedar siding that contrasts with the landscape. Fiber-cement panels with bold colors accent the natural wood finishes of the elevations.

This family retreat uses powerful forms and evocative materials to embrace the natural world.

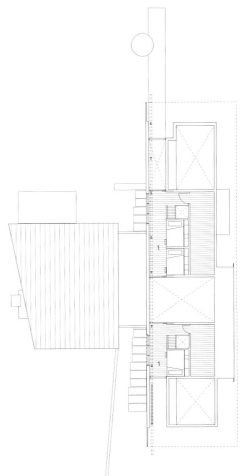

1 双层床卧室　1. bunk room
阁楼　loft

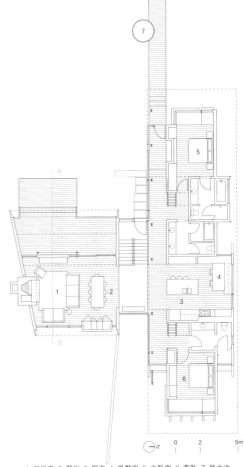

1 起居室　2 餐厅　3 厨房　4 早餐室　5 主卧室　6 客卧　7 热水池
1. living room  2. dining  3. kitchen  4. breakfast room
5. master bedroom  6. guest bedroom  7. hot tub
一层　ground floor

# 布鲁恩别墅
Häkli Architects

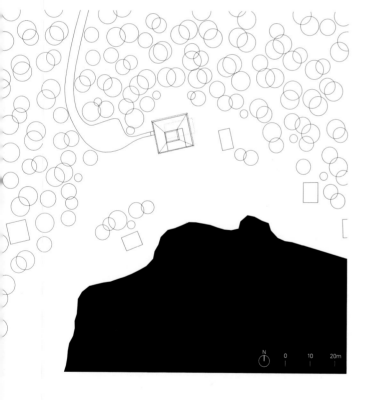

**项目名称**: Villa Bruun
**地点**: Puutossalmi, Kuopio, Finland
**建筑师**: Seppo Häkli
**项目团队**: Hanna-Maija Matikainen
**结构工程师**: Eelon Lappalainen / **HVAC**: Markku Kallio
**电气工程师**: Heikki Pekonen
**承包商**: Jari Koponen, Matti Antikainen
**用地面积**: 4,200m² / **总建筑面积**: 130m²
**竣工时间**: 2012
**摄影师**: ©Jussi Tiainen(courtesy of the architect) - p.112
©James Silverman - p.106~107, 108~109, 110, 111, 113

充分利用日光是设计这座四季休闲住宅的出发点。建筑四壁高高的窗户使日光在白天照进室内的中心,并且成为房子里日常生活的一部分。

建筑包围着中央庭院且面对着景色优美的卡拉韦西景观。带有嵌入的壁炉的客厅和卧室构成了整座建筑的核心,壁炉令这座住宅在冬日里依然像夏天一样温暖。其他房间可在需要时备用。

除了砖瓦面的地面,整座住宅均用木材建造。内外墙壁都采用厚实、优质的企口松木板构成,木板选用大量心材,使颜色多种多样。无框的门窗也由松木制成。

建筑的外表面会随着时间的流逝而渐渐褪成灰色,因而,设计师选用透明的木材保护剂来减少颜色的褪变。室内的门和家具都涂以木油。倾斜的天花板精选无结的白杨木来建造,这种木材能够将高窗入射的日光反射出来。整体配套家具都采用直交层合松木建造,嵌入室内。门和通风窗为手工制造而成。

这座建筑由一个技艺精湛的木匠建造完成,最终体现出其细木工制品的质量。设计将平面和剖面融合为一体。建筑过程中没用上一张图纸,每当建筑工人需要设计细节时,只需直接在建筑师的电脑文件中查看就可以了。

## Villa Bruun

Making the most of the daylight was the starting point for the design of this all-year-round leisure home. The high-level windows encircling the building bring the daily sun cycle into the heart of the interior and make it a part of everyday living.

The building wraps round an atrium facing the Kallavesi Landscape. The living room and bedroom with their built-in fireplace form the core of the building, which can be kept warm in winter as

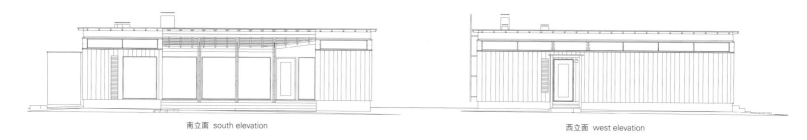

南立面 south elevation　　　　　　　　　　　　　　　西立面 west elevation

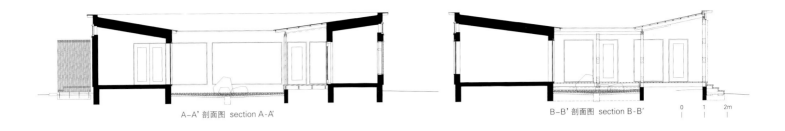

A-A' 剖面图 section A-A'　　　　　　　　　　　　　　B-B' 剖面图 section B-B'

well as summer. Other rooms are used when required.

Apart from the brick-tile finished floor, the entire construction is in timber. Internal and external walls are in thick, top-quality, tongued-and-grooved pine boarding with plenty of heartwood to provide variation in colour. The doors and windows, fixed without architraves, are also in pine.

The external surfaces will turn grey over time and the colourless wood-preservative will reduce colour variation. Internal doors and furniture are treated with wood oil. The slanting ceilings are in selected, knot-free aspen which reflects the light coming in through the high-level windows. The fixed furniture was built-in from cross-laminated pine. Doors and ventilators were all hand-made. The building work was done with exceptional skill by a single carpenter and the final result is of joinery quality. The building designs integrated the architect's plans and sections. No paper was needed – when the builder needed a detail, he looked for it directly from the architect's computer files.

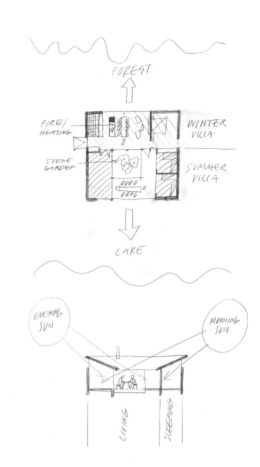

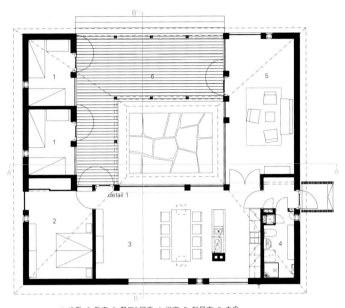

1 次卧 2 卧室 3 餐厅&厨房 4 浴室 5 起居室 6 中庭
1. sleeping cabin  2. sleeping room  3. dining & kitchen  4. bathroom  5. living  6. atrium
一层 first floor

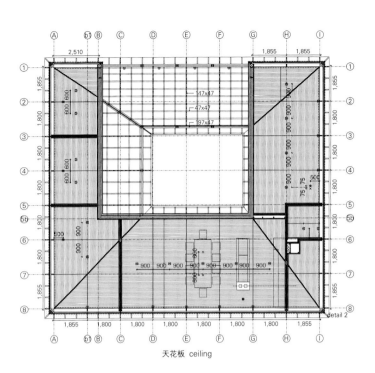

天花板 ceiling

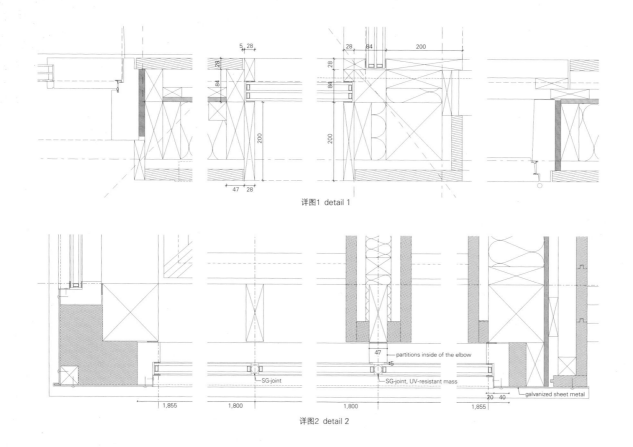

详图1 detail 1

详图2 detail 2

- partitions inside of the elbow
- SG-joint
- SG-joint, UV-resistant mass
- galvanized sheet metal

场地中央的几棵长成的大树（一棵红桉树、两棵美叶桉）在该项目的外观设计中发挥了重要作用。建筑师设计的首要部分是研究是否需要保留这些树木，并利用这几棵树在这样一个毫无特征的普通场地所突出的独特性来说服甲方。在知名树艺师的帮助下，双方观点最终达成一致，保留这些树木。因而，这座住宅就建在树干之间，而其外形则塑造了两座形状不规则的开放式庭院。这些建筑部分将树木和周围的景观环绕起来，甲方便在其间享受家庭生活。

餐厅高高的窗户和主卧室中潜望镜般的棚屋成为这几棵树木在屋内的忠实拥护者，甲方可以在这里看到树叶和脱皮的树干。这几棵树的根系和随风摇摆的巨大树枝给这座住宅的建造带来了挑战。建筑师们更愿意将该项目看作是一个互惠互利的规划设计：将房子建在哪里，才能既保留这些树木，又能通过这些树在视觉上衬托出房子内部空间的质量。

建筑师采用了钢质三角基脚矩阵，来保护树木浅层根系的完整性，每个基脚都需要徒手挖掘安置，而且每次碰触到树木的根系，都要重新改变位置。这样，这就造成结构网格的不规则性。这些基脚也将房子抬离地面，或多或少给人一种临时住宅的感觉。

这座住宅在地面留下的任何与树木根系相重叠的建造印记都会导致雨水不均匀地灌溉这些根系，这对树木来说简直是巨大的冲击。因此，建筑师将屋顶汇集的雨水引到房子下面，通过细细的灌溉管道平稳地浇灌树木的根系。

轻质的构造看似对满足现有树木的要求来说是最合适不过的了，而草砖却成为墙壁保温的优选构造。这个决定的做出就要求所有围墙都要预先建造成梯形架构，并现场搭砌。因此，最后砌成的围墙特别厚，这在木质构造的建筑中很少见。

这个住宅的主要建筑材料是木材，内外空间都采用木材构造，形成纹理。墙衬采用不同等级的胶合板建造，其中，外墙采用粗锯纹理，喷以黑漆，而内墙则采用喷砂处理，清漆。地板和天花板也采用透明喷漆的胶合板。屋顶用单板层积材横梁构造，直接暴露在天花板内。

## Karri Loop House

The mature trees located in the middle of the site(a Karri and two Marris) played an essential part in shaping project. The first part of the design process was spent in investigating the requirements for retaining these trees, as well as convincing the clients of their unique presence on an otherwise anonymous site. With the support of a renowned arborist, the decision was finally made to keep the trees. As a result, the house sits in between the tree-trunks and its outline defines two open courtyards of irregular shape. These embrace the trees and the surrounding landscape, around which family life occurs.

A tall window in the dining area and a periscope-like skillion in the master bedroom, celebrate the presence of the trees from within the house, framing views of both foliage and peeling trunks.

# 环绕红桉的住宅
MORQ

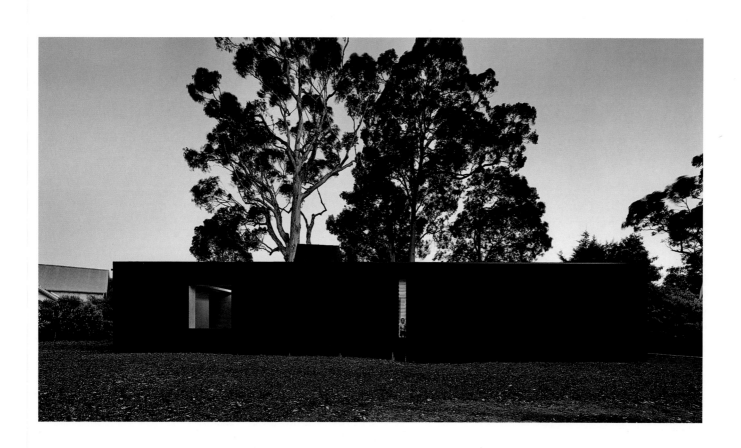

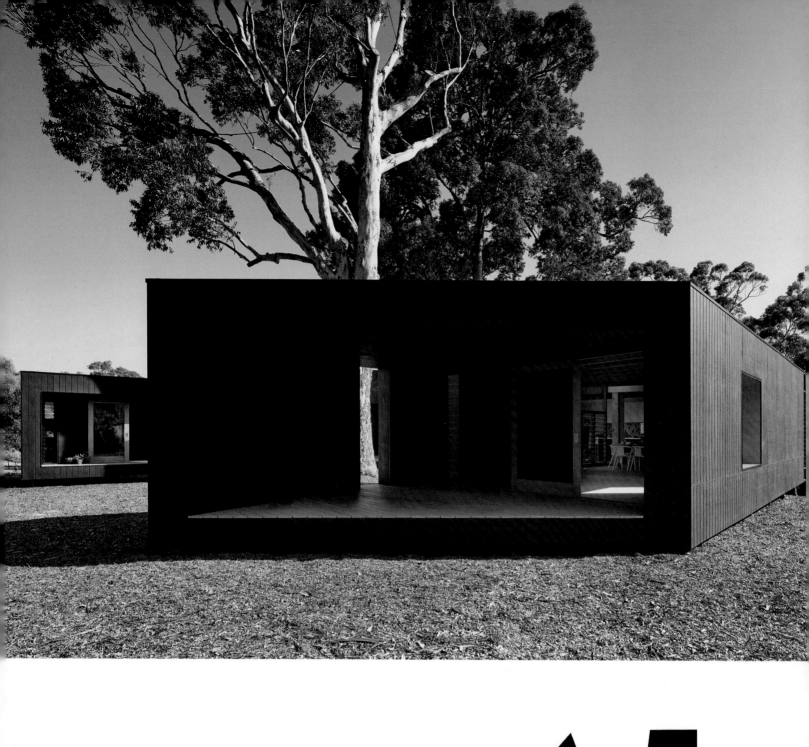
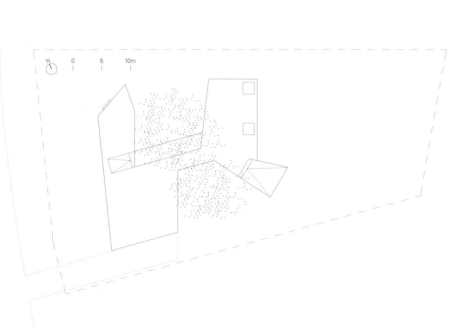
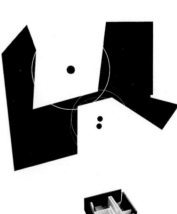
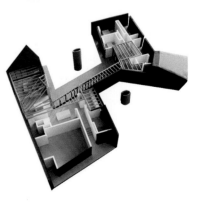

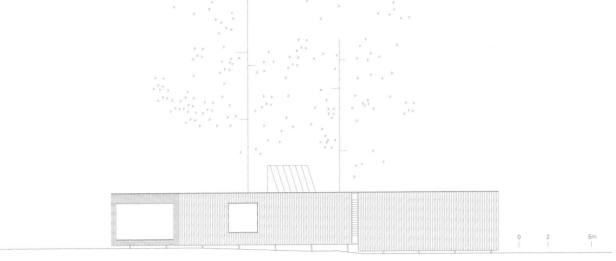

西立面 west elevation

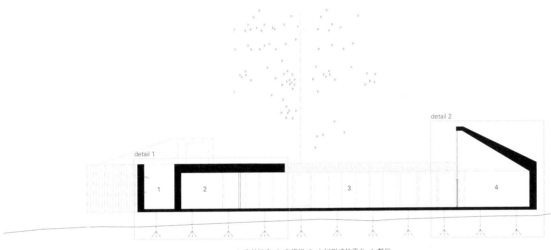

1 室外浴室 2 电视间 3 大树形成的平台 4 餐厅
1. outdoor bathroom  2. TV room  3. tree deck  4. dining
A-A' 剖面图  section A-A'

1. entry 2. garage 3. laundry 4. powder room 5. kitchen 6. dining 7. living 8. covered deck 9. tree deck
10. study 11. TV room 12. master bedroom 13. master en suite 14. outdoor bathroom 15. children's bedroom 16. bathroom

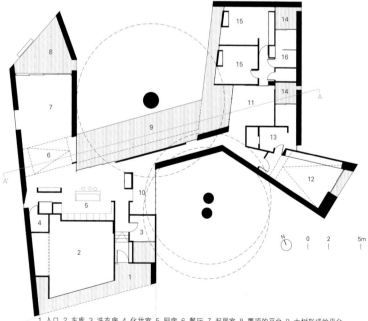

项目名称：Karri Loop House
地点：Margaret River, Western Australia
建筑师：MORQ
合作者：Josh Saunders, Tor Dahl, Ken Yeung, Catherine Farrell, Clare Porter, Sally Farrah
建造商：Tectonics
结构工程师：Margaret River Structural Engineers
首席木匠：Henk Van Oostenbrugge
用地面积：290m² / 有效楼层面积：300m²
设计时间：2009
施工时间：2013
竣工时间：2013
摄影师：©Peter Bennetts

These trees, their root systems and their unstable large branches presented a challenge to the build-ability of the house. The architects like to think of this project as a mutually beneficial development: where the building is designed to retain the trees, while the trees visually contribute to the quality of the inner space.

To protect the integrity of the shallow root-system a matrix of steel tripod footings was used: each of them had to be dug by hand, and repositioned every time when a root was encountered, resulting in an irregular structural grid. These footings also raise the house off the ground and give it a somewhat temporary look. Any part of the house footprint overlapping the root system would result in an uneven rainwater supply to the roots, which could cause a shock to the trees. Rainwater collected on the roof is therefore taken under the house, channelled into trickling irrigation pipes and then evenly fed to the tree roots.

Lightweight construction seemed to be the most appropriate response to the existing trees requirements, however straw-bales were chosen as a preferred form of insulation. This decision required all perimeter walls to be prefabricated as ladder-frames and later assembled on site. It also resulted in unusually thick perimeter walls, seldom employed in timber framed buildings.

The house was mainly constructed out of timber, whose grain and texture inform both interior and exterior spaces. Wall linings use different grades of plywood: rough sawn, painted black on the outside, and sanded, clear-treated on the inside. The floor and ceilings are also in clear-treated plywood. The roof structure is resolved with Laminated Veneer Lumber beams, which are left exposed on the inside of the ceiling.

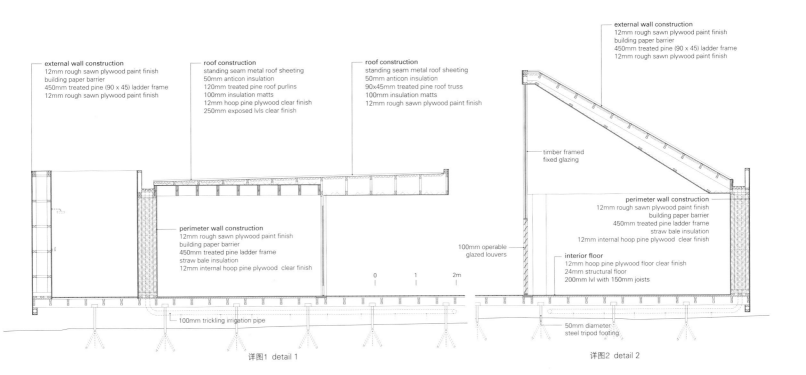

详图1 detail 1

详图2 detail 2

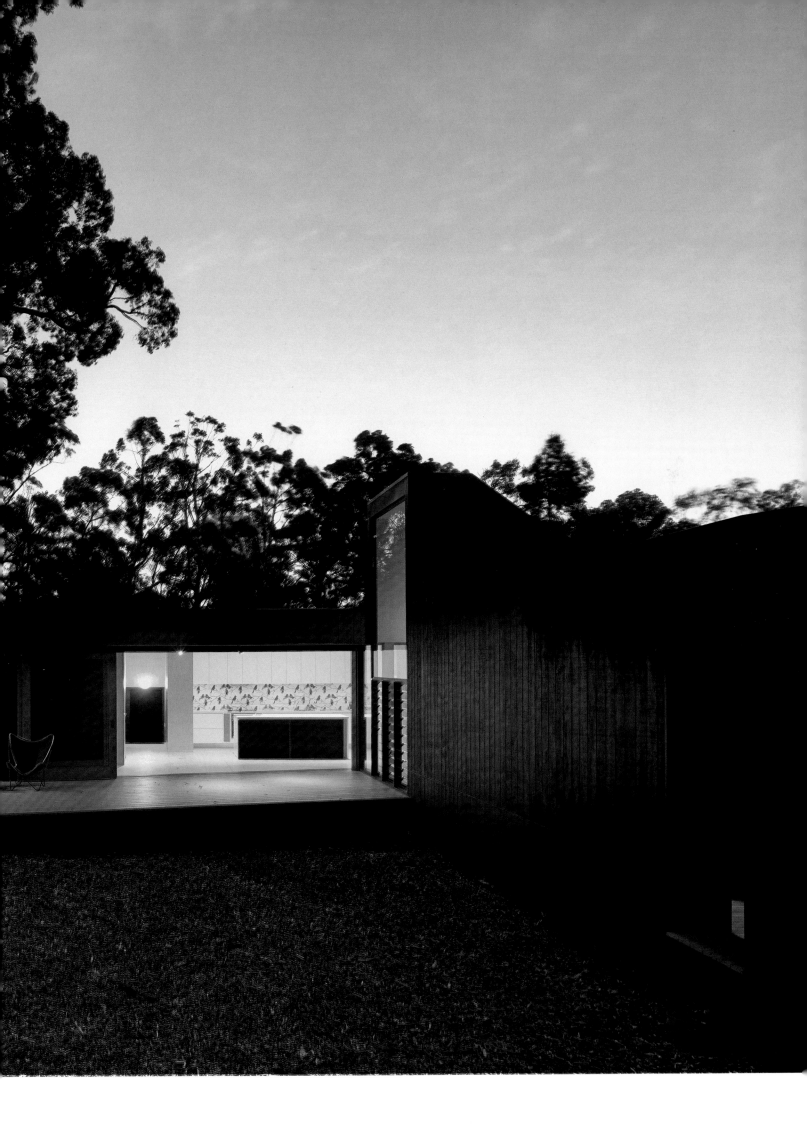

# 卡萨格兰实验室

# Casagrande Laboratory

马可·卡萨格兰是卡萨格兰实验室的负责人,他是一位芬兰籍的建筑师、环境艺术家,以及社会理论家。自他的职业生涯早期开始,他就开始将建筑与其他和艺术科学相关的学科结合起来。从1999年开始,他创作了65件跨领域的、原创的、激进的建筑作品。

他的作品和教学在建筑、景观、环境艺术、城市与环境设计和科学间自由穿梭,以增加跨学科的建筑思想,具有建筑人文环境的广阔视角。

C3采访了马可·卡萨格兰,听了关于他的作品的很多故事,也领略了其关于"自然""地域""再生""工作方法""建筑师角色以及建筑领域"的思想。

Marco Casagrande, the Principal of the Casagrande Laboratory, is a Finnish architect, environmental artist, and social theorist. From the early stages of his career, he started to mix architecture with other disciplines related to art and science. Starting from 1999, he has created 65 cross-disciplinary, original and radical works.

His works and teaching are moving freely in-between architecture, landscape, environmental art, urban and environmental design and science adding up into cross-over architectural thinking, a broad vision of built human environment.

C3 has interviewed Marco Casagrande to hear more stories about his work and thoughts on "Nature", "Locality", "Regeneration", "Method of Work", "Role of Architect and Realm of Architecture".

## 与自然共鸣
## Resonating with Nature

C3：根据您作品设计的广阔范围（包括建筑、建筑设施、环境艺术品等），您看起来不止对和建筑相关的很多事情感兴趣。有什么特别原因吗？

马可·卡萨格兰：建筑都是开放的形式。建筑的灵感激发于它本身与其他元素的关系，像是一个空间工具。建筑本身不能独立存在。当建筑与自然产生共鸣时，就像鸟的歌声让我们触动。我想要听到这种歌声，但是如果建筑的形式是关闭的，我就听不到了。歌声和鸟鸣背后的共鸣可以通过其他的艺术形式感知，也可以利用真正的工程和真正的科学来感知……都与自然产生共鸣。我对这种共鸣很感兴趣，不用考虑与其相关的艺术和科学手段。

C3：自然看起来是您很多作品的灵感，并且是您生活的固有部分。您能描述一下您的设计中自然所占的位置吗？

| | |
|---|---|
| 台东废墟学院 | Taitung Ruin Academy |
| 台北废墟学院 | Taipei Ruin Academy |
| 终极废墟 | Ultra-Ruin |
| 财富山 | Treasure Hill |
| 波将金公园 | Potemkin |
| 陈宅 | Chen House |
| 昆虫屋 | Bug Dome |
| 乌尼岛夏季剧院 | Uunisaari Summer Theater |
| 景观 | Land(e)scape |
| 牡蛎人 | Oystermen |
| Paracity项目 | Paracity |

**C3** According to your broad work range (including architecture, architectural installation, environmental artwork, etc.), you seem to be interested in issues that are not only related to architecture. Is there any special reason for this?

**Marco Casagrande** Architecture is an open form. It is activated by its relationship with other elements, like a spatial instrument. Architecture alone does not exist. When architecture resonates together with nature, it touches us like the singing of the birds. I want to hear this singing, but I cannot hear it when the form is closed. This same song and resonation behind the birds' singing can be felt in other forms of art as well, as well as with real engineering or real science […] all resonating with nature. I am interested in this resonation, regardless of the artistic or scientific means used to connect with it.

**C3** Nature seems to have been the inspiration for so much of your work and it seems to be an intrinsic part of your life. Could you describe the position of nature in your design?

**MC** Design cannot replace reality – namely, nature. Architectural

MC：设计不能取代现实，名义上的自然。建筑控制必须是开放的，容许自然加入进来。

这就是"弱"建筑的意思。同时城市必须被毁掉，以成为自然的一部分。人类的有些管理是污染的来源。建筑不应该是管理的艺术，相反它是现实的艺术：即自然。观念要求我们成为建筑的一部分。如果没有这些，人类就需要治疗了。如果建筑人文环境将我们与自然分离，建筑就该被毁掉。城市充满着治疗需求。自然可以让建筑工具唱起歌来——它让我们全都唱起来——甚至城市也唱了起来。

昨天我儿子两岁了。我们睡在满是雪的森林里来庆祝生日。晚上，他在黑暗的森林里盯着篝火，周围都是满是积雪的树，他像个萨满法师一样对着篝火唱歌。

我在芬兰北部的拉普兰长大。我的童年记忆都和自然有关。我想将这种联系贯穿在我的作品中。我也可以为其他人建立这种联系。

## 地域性

C3：您在全世界都工作过。您能解释一下您是如何使用不同的方法将自己置身于不同的环境中的吗？

MC：我对当地的文化知识很感兴趣，并且很敏感地将它阐释出来。当地的知识远远超出文化，它是一种在既定的环境中如何和自然相处的特殊智慧。有时我们说文化是自然和人造自然之间的阻碍力量。当地的知识可以将其连接起来。所以，它在现存的城市范围外呈现出描绘未来的工具。当地知识的元素是第三代城市成长的种子。这些种子并非一定是正式的，但是一定在城市最脏的地区孕育，在城市的肥沃肥料中，最后被呈现出来，进而展现出肥沃的表土，利于城市有机物的发展。

Paracity项目是个很好的例子。简单地说，它是一个木质基本结构的模型，城市有机体可以建于其中，这里通常是那些工业城市不愿意去发展的地方：城市河流的灾区平原、海啸易侵袭的地方和贫民窟。我们只提供基本的结构，即城市的骨架，在此基础上人们建造自己的家园、社

control must be opened up in order to allow nature to step in. This is what "weak architecture" means. Also cities must be ruined so that they can become part of nature. Human control is the source of pollution. Architecture should not be the art of control; rather, it is the art of reality: nature. Consciousness means us being part of nature. If this doesn't happen, man needs therapy. If the built human environment separates us from nature, it should be destroyed. Cities are full of therapy. Nature makes the architectural instrument sing…it makes us all sing – even a city.

Yesterday my son turned 2 years old. We celebrated his birthday by sleeping in the snowy forest. At night, he was staring at our fire in the dark forest, with big trees full of snow all around him, and he was singing to the fire like a shaman.

I grew up in Lapland, in northern Finland. All my childhood memories are connected with nature. I want this connection to continue happening through my works. I can build this connection for other people.

## Locality

C3  You have been working globally. Could you explain your approach to each project that places you in very different environments?

MC  I am interested in the local knowledge and have some sensitivity to interpret it. Local knowledge is beyond culture; it is the site-specific wisdom of how to live together with nature at a given site. Sometimes what we call culture can be a stopping power between nature and human nature. Local knowledge has this connection and, therefore, it presents tools for futures beyond the current city. The elements of local knowledge are the seeds of the third-generation city. These seeds are not necessarily official, but being composted in the smelliest parts of the city, in the urban composts, and are to be turned over in order to present the most fertile top-soil for the organic urban development.

The paracity is a good example. Basically, it is a modular wooden primary structure for a city organism that can grow in places where the industrial city is not likely to go: flood plains of urban rivers, tsunami-prone areas and slums. We only provide the primary structure, the skeleton of the city onto which people will realize their own homes, communities, and urban farms. The paracity will always be local and brought to life from the local culture and knowledge.

区和城市农场。Paracity项目永远是当地的，因为当地的文化和知识而被赋予了生命。

C3：您最近在亚洲的工作（尤其在中国台湾和日本）非常引人瞩目。您认为亚洲的特征是什么？您认为亚洲和欧洲有什么区别吗？

MC：亚洲和欧洲是同一块大陆的不同角落。我可以从芬兰走到韩国。我们完全不同。自然将我们塑造得如此不同。但基本的元素是一样的，建筑也是如此，因为建筑是人类的一个基本的表达。一个人不应该只停留在事物的文化层面上，应该感受鸟鸣背后的共鸣。亚洲自然——肥沃的溪谷和冲积平原——有利于相对原始的城市生活，但这在芬兰的自然环境中是不可行的。我们生活在森林里。拉普兰环境更差，森林都很少，多是冻土，甚至村庄都无法存留。人们必须迁徙。

第一代亚洲城市依旧是人性作为自然一部分的表达，这种特质在作为当地知识体现的现代城市中仍旧留存。现在机器工业化城市从自然中脱离，并违背自然规律，但当地知识依旧有所体现，我们必须学会使用，用于生物化城市的复兴中。

**复兴**

C3：废旧场地和建筑的复兴是您作品的大部分。对于这类作品您有自己的方法吗？您能和我们介绍下吗？

MC：废墟都是开放的形式。一些人工的东西已经成为自然的一部分。看到自然读懂建筑是很有希望的事情。在台北，有一颗树———个城市的盆景——长于公寓的墙上。树有五层楼高，根也不在地上。树将房子作为它生长的平台和生命的来源。它的根一部分深入到下水道和房子的排水系统中，吸取能量；另一部分深入到建筑的钢筋混凝土板材中，成为结构。树深知人们卫生间的使用频率以及在这个基础上它可以长多大。一个月前，当我向帕拉斯莫教授介绍这棵树时，他问："人们应该写本关于这棵树的书。对于决定读懂建筑方式的这棵树来说，它具有怎样

---

**C3** Your recent work in Asia (especially in Taiwan of China and Japan) is noteworthy. What do you think are the characteristics of Asia? And what would be the distinction between Asia and Europe?

**MC** Asia and Europe are two different corners of the same big continent. I can walk from Finland to Korea. We are all different. Nature has shaped us differently. Fundamental things are the same; the same is true to architecture, which is a fundamental human expression. One should not stop looking at only the cultural layers of the things, but feel the resonance behind the singing of the birds. Asian nature – its fertile river valleys and flood plains – can support original urbanism, which in the nature of Finland would be fictive. We are forest people. Lapland is even harder: It has no more forest, just tundra, which cannot support even a village. Man must move.

The first-generation Asian cities are still expressions of human nature as a part of nature, and this quality has remained embedded in the contemporary cities as forms of local knowledge. Now the industrial-mechanical cities are developing away from and against nature, but local knowledge is still surviving, and we must learn to use it in the biourban restoration of our cities.

**Regeneration**

**C3** The regeneration of abandoned sites and buildings comprises a large proportion of your works. If you have your own approach to this kind of work, please introduce it to us.

**MC** Ruins are an open form. Something man-made has become part of nature. It is hopeful to see how nature reads architecture. In Taipei, there is a tree – an urban bonsai – that grows from the wall of an apartment building. The tree is 5 stories high, and its roots don't touch the ground. The tree uses the house as its living platform and source of life. Its roots partly go into the sewage and water systems of the house for energy and partly penetrate into the reinforced concrete slabs for structure. The tree knows how much the people are using the toilets and how big it can grow based on the structure of the house. A month ago, when I introduced him to the tree, Professor Pallasmaa asked, "One should write a book about this tree. What is the sense of this tree that determines how it can understand architecture?"

The sensitivity of the ruins comes from opening up the human

的意义？

废墟的感受性来自于人类管理的开放性，所以自然可以介入其中。我们能感受自然的动态性，且我们的人性也可以介入其中。

C3：最近，人们对"城市复兴""可持续发展""小建筑"和"弱建筑"很感兴趣。您对这些主题有什么想法吗？

MC：这些都是对城市集中建造人文环境以及机械工业化城市结构的反抗。工业化城市不能再为市民提供更好的生活。相反，城市和自然的疏远导致人们从城市内部寻找原始的资源，形成了自己的社区和自己的建筑反馈。人们不再信任城市，也不再信任工业发展。城市导致了工业污染，将人类的居住地变成了反对在自然环境居住表面进行"针灸"的"针"。生物化城市的复兴，已经由很多常人得以实现，通过不同的方式在城市的表面进行"针灸"，把城市变成有机体。我们正在进行的关于Paracity项目的调查建立在对模型化、自我组织的城市社区的调查基础上，这些社区探求在中央化的城市中间得以生存和发展。Paracity项目从非官方的社区和当地知识中学习和得到灵感。我们只是提供基本的模型结构，促进生物化城市的发展，在这些模型上人们可以自主建房子、社区和农场。现存城市的生物化复兴是今天人类发展的关键问题。城市成为远离自然的卫星，建筑也同样如此。

**工作方法**

C3：您看起来是使用模拟的方法而不是数字的方法来创作作品。图纸都是类似手绘的素描。这有什么特殊原因吗？

MC：绘图接近于沉思。一个人可以更接近真实的事情。批判性的绘画更有挑战性。我们只是等待绘画出现的奴隶。

C3：您能介绍一下您的实验室吗？您是怎么进行团队合作的？

MC：我们的事务所是一股很小但很有效的跨学科调查和设计力

control so that nature can step in. We feel this natural dynamic, and our own human nature wants to step in as well.

C3 Recently, people are becoming more interested in "urban regeneration", "sustainability", "small architecture" and "weak architecture." Do you have any comment on these issues?

MC These are reactions against the centrally governed built human environment and the structures of mechanical-industrial cities that are no longer providing a better life for their citizens; on the contrary, the urban alienation from nature causes people to find the original resources from within themselves, forming their own communities and their own architectural reactions. They don't trust the city anymore, and they don't trust the industrial development. It is the official city that is causing industrial pollution and turning the human urban settlements into anti-acupuncture needles on the living surface of our natural environment. Biourban restoration has already been realized by normal people in various ways performing acupuncture on the cities, turning them toward the organic. Our ongoing research on paracities is based on the research of modular, self-organized urban communities that seek to survive and grow outside the centralized system. The paracity is learning from and inspired by unofficial communities and local knowledge. We are only providing the primary modular structure for the biourban growth onto which people can independently attach their homes, communities, and farms. The biourban restoration of existing cities is the key question for human development today. Cities are becoming satellites disconnected from nature. The same is happening with architecture.

**Method of Work**

C3 You seem to prefer an analog method rather than a digital method when developing works. Even drawings are hand-drawn close to sketch. Is there any special reason for this?

MC Drawing is close to meditation. One gets closer to the real things. Critical drawing also challenges me. We are just slaves waiting for the drawings to emerge.

C3 Could you introduce your laboratory? How do you cooperate with the team?

MC Our laboratory is a small and effective multidisciplinary research and design force. We have common strengths and individual specialties. I can parachute these guys anywhere in the world

量。我们有共同的力量和个人的特性。我可以将他们空降到世界的任何地方，然后给他们任务，像特种部队。不久，我们会有C-130大力神运输机，把我们的办公室搬到那里。

**建筑领域和建筑师角色**

C3：您对建筑的界限有何想法？

MC：建筑没有界限。建筑像细胞质。它能从很多事情中吸收和得到能量——它是大脑的一部分。首先人们有东西要说，然后找到一种表达方式。自然和建筑师是合作关系，不只是我们在建造。

C3：您的建筑总是在传达特殊的信息。通过看自然、环境、城市和人类，我们开始思考生命的基本价值。您认为建筑师的角色是什么？

MC：某种意义上，建筑师是利用敏感度来诠释集体智慧是怎样利用建筑来进行传输和反馈的人。集体智慧也是和自然联系的，自然就是个大型神经系统。集体智慧的表达是个复杂的过程，不能通过试图控制它来达到。控制是死亡的伙伴，不是生命的必要表达。建筑师是一种设计类的萨满法师。每个实际的项目都要置之死地而后生。

C3：您的建筑事务所如何处理未来这一概念？

MC：我们是在未来。我们如何处理现在才是最关键的问题。Paracity项目是我们处理现在、过去、未来的工具。它会成为活的城市有机体，与其说是城市，还不如说是个神经系统。

C3：我们认为您远远不只是个建筑师，您如何定义自己？

MC：谢谢您，但是建筑师的头衔对我来说便足够了。

"波库派恩发生了什么事情？"

"某天他从Zone回来，他变得非常富有，非常富有。第二周他自杀了。"

and give them a mission, like Special Forces. Pretty soon, we will get a C-130 Hercules and move our offices there.

**Realm of Architecture and Role of Architect**

C3  What do you think about the boundaries of architecture?

MC  There are no boundaries. Architecture is more like plasma. It can absorb and get energy from many things – it is part of the big brain. First one has to have something to say and then find a way to say it.

Nature should be understood as the co-architect. We are not building alone.

C3  Your work always delivers a special message. Through looking at nature, environment, city, and human-being, we begin to think about the fundamental value of life. What do you think about the role of architect?

MC  In a sense, architects are the people who have the sensitivity to interpret how the collective mind is transmitting and reacting to it with construction. The collective mind is also well connected with nature, which is basically one big nervous system. The articulation of the collective mind is a highly complex task that cannot be achieved by trying to control it. Control is death's companion, not necessarily an expression of life . Architects are a kind of design shamans. One has to die a bit to be reborn in every real project.

C3  How does your practice intend to address the future?

MC  We are in the future. How can we address the present is the more crucial question. The paracity is our tool for working with the present, past, and future. This can become a living organic urbanism, more like a nervous system than a city.

C3  We think that you could be called "more than an architect." Could you define yourself?

MC  Thank you, but architect is plenty.

*"What really happened to the Porcupine?"*

*"One day he came back from the Zone and he got amazingly rich … amazingly rich. The next week he hanged himself."*

废墟学院建立在一个位于中国台湾台东的废弃制糖厂（1913~1996）内，其目的是进一步对第三代城市理念进行城市生态研究。废墟学院的核心位于工厂的蒸汽箱区域，从这里开始，这个区域将逐渐占据整个工厂的全部面积。它也为废墟学院的工作间和工作室（大面积的社区园艺和厂内野生自然环境为其提供基础）提供了多方面研究和教育空间。花园上方的屋顶有几处被拆除，这是为了方便雨水进入。另外，原来的屋顶以及雨水通道都用来将水源储存到水箱之中，再用于灌溉。所有的地板都是由泥土或者木材铺成，人们可以赤脚在这个有机的机器内行走。

## 台东废墟学院

台东废墟学院致力于研究台东县地区台湾原著民部落的智慧，而后将这种智慧融入生态城市（现有城市的生态恢复）的设计理念中。原著民的生活仍然与自然有着亲密的联系，他们是抵制现代工业的。这种人与自然之间的生态联系将作为第三代城市理念的决定性因素再次被引入工业城市中，即工业城市内的有机废墟，也是一架生态机器。在一切都人工化之后，废墟便成为自然的一部分。

废墟学院的中央是一处灌木丛，在废弃工厂的机器中肆意生长。所有学术设施、榻榻米房间、多功能空间、未来将建造的长廊和礼堂、桑拿中心等，都在一簇簇丛林中应运而生。

此废墟学院的研究中心、工作间、工作室的成立是各行各业的台湾人以及各个国际大学合作的结果。废墟学院成为国际生态城市社会的一部分。

## Taitung Ruin Academy

The Ruin Academy is set up in an abandoned sugar factory (1913~1996) in Taitung, Taiwan, China in order to take further bi-ourbanist research of the Third Generation City. The core of the Ruin Academy is located in the evaporator tanks area of the factory from which it will gradually grow to occupy the whole sugar factory. The core offers multifunctional research and educational spaces for Ruin Academy workshops and studios backed up by extensive community gardening and wild nature inside the factory. The roof above the gardens has been removed in selective points in order to let rain inside. Additionally the existing roofs and rain water channels are being used for harvesting rainwater into water tanks for irrigation. All the floors are covered by wood or earth – one can walk bare-foot inside the organic machine.

The Taitung Ruin Academy is focused in the research of the Local Knowledge of the indigenous Formosan tribes in the Taitung County area and transforming this organic knowledge into design methodologies of biourbanism – ecological restoration of existing cities. The indigenous communities are still living in a continu-

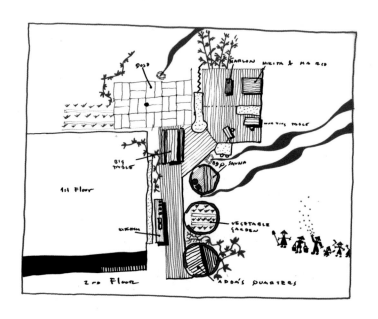

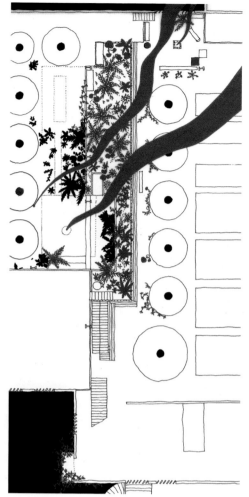

三层 third floor

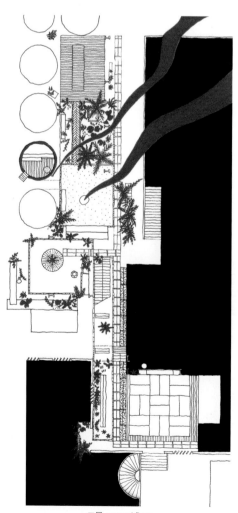

二层 second floor

项目名称：Taitung Ruin Academy
地点：Taitung Sugar Factory, Taitung, Taiwan, China
建筑师：Marco Casagrande
管理者：Sandy Hsiu-chih Lo
项目经理：Nikita Wu
车间作业员：AdDa Zei(C-LAB) / Vaan Wang, Mao Hui, Shuai Wen-Hao / Yang Shan-Chun / Mao Ching-Yi, Li Ying-Ying / Wang I-Hsiang / Yu Chia-Jen / Lee Me Fe / Hsu Chu-Chun / Chen Ruo-Qing / Lai Yu-An / Niu Kuang-Yu, Guo Ren-Chiao, Huang Yu-Nan, Chung Wen-Hao, Yen Yu-Tien / Peng Qien Ru
当地顾问：
Curator of Taitung Sugar Factory Museum _ Tian-Cai Lin / Horticulturalist _ Mei-Hsiu Wang
甲方：Taitung County Government
支援者：Deputy governor _ Chang Chi-Yi
楼层面积：450m²
材料：wood, soil, brick, white stone, tatami, recycled industrial elements
竣工时间：2014

©Hope Monkey (courtesy of the architect) - p.132 first, p.136 top
©AdDa Zei (courtesy of the architect) - p.130, p.131, p.132 second, third, fourth, p.133, p.134, p.135, p.136 bottom, p.137

ous close connection with nature and are resisting the industrial control of the modern man. This biological connection of man and nature will be re-introduced to the industrial city as a decisive step towards the Third Generation City, the organic ruin of the industrial city, an organic machine. Ruin is when man-made has become part of nature.

The heart of the ruin academy is a jungle growing from the machinery of the abandoned factory. All the academic functions, tatami room and multi-functional spaces, future galleries and auditoriums, and the public sauna are the fruits of this jungle.

The Ruin Academy research, workshop and studios are realized in co-operation with various Taiwanese and international universities. Ruin Academy is operating as a part of the International Society of Biourbanism.

## 台北废墟学院

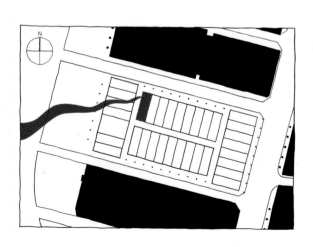

## Taipei Ruin Academy

Taipei Ruin Academy is an independent cross-over architectural research center in the Urban Core area of Taipei, Taiwan, China. The Academy is run in co-operation between the Finland based Casagrande Laboratory and Taiwanese JUT Foundation for Arts & Architecture.

Ruin academy is set to re-think the industrial city and the modern man in a box. It organizes workshops and courses for various Taiwanese and international universities. The research and design tasks move freely in-between architecture, urban design, environmental art and other disciplines of art and science within the general framework of built human environment.

The Ruin Academy occupies an abandoned 5-story apartment building in central Taipei. The interior walls of the building and all the windows are removed in order to grow bamboo and vegetables inside the house. The professors and students are sleeping and working in mahogany made ad-hoc dormitories and have a public sauna on the 5th floor. The building is penetrated with 6 inch holes in order to let "rain inside". The Academy is viewed as an example or fragment of the Third Generation City, the organic ruin of the industrial city.

The Ruin Academy does not rely on design, but hooks on to the Local Knowledge of the Taipei basin and reacts on this. Design

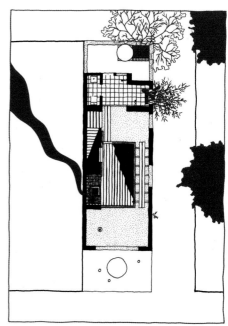

五层 fifth floor

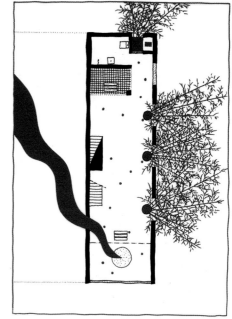

四层 fourth floor

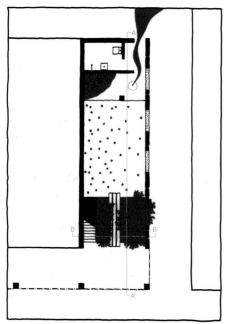

一层 first floor

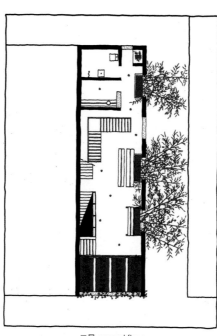

二层 second floor

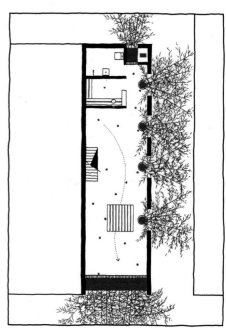

三层 third floor

should not replace reality. Local knowledge is pushing through the industrial surface of the modern Taipei. The Ruin Academy is looking at the ruining processes of Taipei that keep the city alive. The Ruin Academy operates with Taipei as the urban case study and with various smaller projects in Taiwan in order to determine the elements of the Third Generation City. The students/operators are not volunteers, and they are called constructor-gardeners. The architects want to farm a city and treat it as urban acupuncture tuning the city towards the organic.

The 3G City is an organic matrix of nature mixed with human construction.

It is light on the surface but with solid roots. The architectural control is in a process of giving up in order to let nature to step in. Modernism is lost and the industrial machine will become organic. This happens in Taipei and this is what the architects study. Ruin Academy is an organic machine. Marco Casagrande

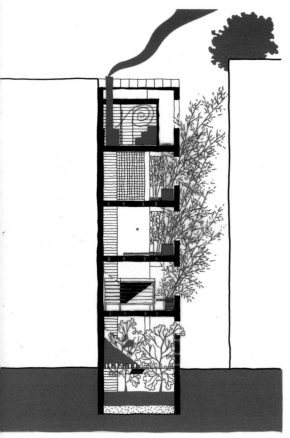

A-A' 剖面图  section A-A'

B-B' 剖面图  section B-B'

项目名称：Ruin Academy
地点：Taipei, Taiwan, China
建筑师：Marco Casagrande
项目经理：Nikita Wu
项目团队：Frank Chen, Yu-Chen Chiu, AdDa Zei
总建筑面积：500m²
总建筑规模：five stories above ground
材料：concrete, mahogany, white gravel, top-soil, 6 inch holes. bamboo, taro, Chinese cabbage, passion fruit, Asplenium nidus, wild trees, ferns and undergrowth．
竣工时间：2010
摄影师：©AdDa Zei(courtesy of the architect)

144

卡萨格兰实验室 Casagrande Laboratory

终极废墟

　　终极废墟是一个木质的建筑有机体,是从一个废弃的红砖农舍废墟发展起来的,位于阶梯状的农场和丛林的交汇处。建筑师遵循开放形式的原则,根据展现的丛林、废墟和本地信息的特点在现场进行创作。

　　该综合体内设有各种各样的多功能空间和平台,它们提供了不同的生活功能和沉思空间。内部和外部的空间连续性不是固定的——内部空间可以是室外,而丛林也仿佛位于建筑内部。终极废墟如同一件由自然(包括人类)进行"演奏"的"乐器"。建筑的主要使用者是一个私人家庭,但其空间偶尔会因为大型会面活动而对外开放。

　　与其说终极废墟以工业控制为基础,倒不如说它是一个有机体。偶然性比建筑控制更加重要。建筑师曾经放开对建筑的管理,让自然进入其中,且允许人类错误的发生。理解偶然事件的动态性,就要把它展现出来。展现是所有艺术的关键元素。

　　建筑并不是一种独立的语言,也不是在自言自语。建筑需要自然,以成为自然的一部分。终极废墟是一处传递废墟感的环境,在这里,人们回到家(废墟)中,并且与丛林同宿。

　　终极废墟自2009年(甚至现在仍在)与甲方进行亲密对话。建筑师的第一个举动便是建造一个人们围绕而谈的桌子,之后便为这个桌子设计一个遮盖棚。终极废墟的其他部分都是围绕这个第一印象而发展起来的。终极废墟一直在发展,成为一个开放的形式。

**Ultra-Ruin**

Ultra-Ruin is a wooden architectural organism that is growing from the ruins of an abandoned red brick farmhouse in the meeting place of terraced farms and jungle. The architecture follows the principles of Open Form and is improvized on the site based on instincts reacting on the presence of jungle, ruin and local knowledge.

The complex has a variety of multi-functional spaces and platforms that can be activated for different living functions and for meditation. The spatial continuity between interior and exterior spaces is flexible – also the inside is out and the jungle is in the house. The Ultra-Ruin is an architectural instrument played by na-

屋顶 roof

项目名称：Ultra-Ruin
地点：Yangming Mountain, Taipei, Taiwan, China
建筑师：Casagrande Laboratory
项目经理：Nikita Wu
项目团队：Marco Casagrande, Frank Chen, Yu-Chen Chiu
总建筑面积：210m²
有效楼层面积：730m²
材料：Mahogany, Zelkova, Camphor, Taiwan Cypress, bronze, steel, stone, brick
设计时间：2009
施工时间：2013
竣工时间：2013
摄影师：©AdDa Zei(courtesy of the architect)

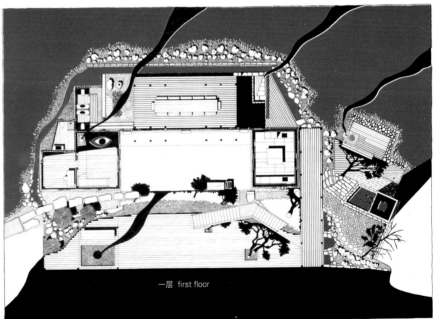

一层 first floor

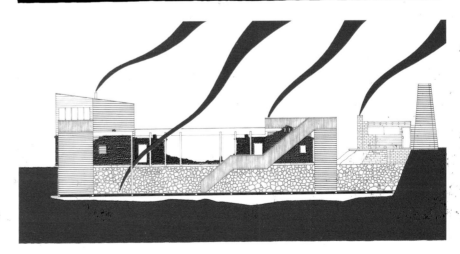

ture including human. The main user is a private family, but the space is occasionally opened up for wider meetings.

Ultra-Ruin is more of an organic accident, than based on industrial control. Accident is greater than architectural control. Architectural control has been opened up in order to let nature to step in and human error to take place. In order to understand the dynamics of an accident one must be present. To be present is the key of all art. Architecture is not an independent language and architecture is not talking alone. Architecture needs nature to become part of nature. Ultra-Ruin is a post-ruin condition, where human has come back to the house/ruin and share the same space with jungle.

The Ultra-Ruin has been developed in close and still ongoing dialog with the client since 2009. The first architectural reaction was to build a table around which people could talk, then to build a shelter for this table. The rest of the Ultra-Ruin has grown up around this initial impact. The Ultra-Ruin keeps growing as an open form. Marco Casagrande

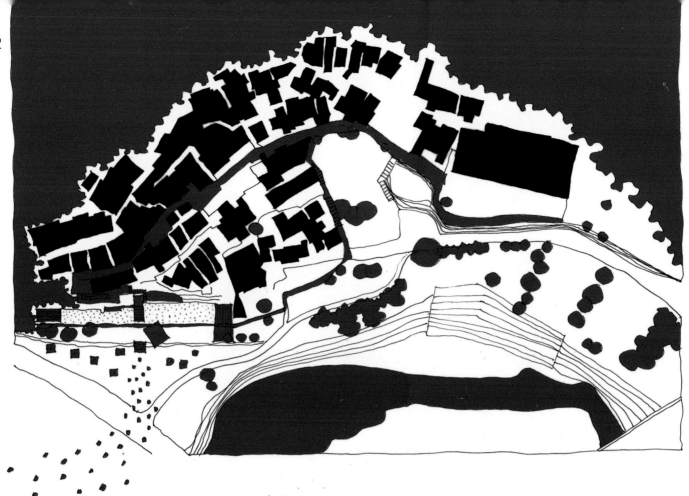

# 财富山

  财富山是一个让国民党老兵的非法定居地变得合法化的地方。他们在这里定居时主要是靠城区农耕,向农民工租赁房屋以及从周围城市收集废品维持生计。

  2002年,我给台北市政府写信,告诉他们这些国民党老兵将会慢慢死去。而我当时主要是从生态学方面考虑这一问题的,也就是说当一个现代人与自然隔绝时,他将会慢慢死去。

  一年后,政府部门给了我回复并且邀请我去台北提供可供选择的生态城市建设计划,以阻止他们死亡。

  我到达那里后,从当局得到了大量的数据,但是它们毫无用处。因为在台北街头的人们与政府提供的信息大相径庭,这些信息更像是一本机器的说明书。街头的人们高尚、活泼并且充满活力。似乎正是他们使城市,这个工业机器,变得鲜活有生气,他们正在一点点地毁灭这架古板的机器,并且努力消除其中的官方色彩。

  我被带到财富山,当时邀请我来台北的市政府正在对其进行拆除,当我到达时,这片连栋房屋的前三排已经被拆除了。所有的城区农田和成片的菜园都被摧毁,为新的市民公园腾出地方。财富山是城区农民的高密度聚居地,而这些农民都是来自大陆的国民党老兵们。

  而令我更为震惊的是这个非法的聚居地有很多值得学习的东西,并且官方政府委托我将其引入现代的台北:这些城区农民用植被净化污水,用符合生物气候学的空气环流来改善被动式太阳房的环境,把有机废物制成肥料,并且尽可能地使用最少的电能,这里没有电视,只有一个集体的小电影院,并且他们从不开车。在这里,城区流浪者们自己种植粮食作物并从被称为荒地的周围城区收获劳动果实。财富山从未有河水决堤的情况出现,因为在这座处于工业化时代的城市,人们早已建造了一个12m高的钢筋混凝土墙,将人类的建造与自然隔开了。

  我改变了我的计划,这样我就可以全身心地去阻止财富山的拆除工作,同时真正地开始重建已被毁坏的房屋之间的连接,并且修复他们的农场。为了完成所有的这些工作,当地居民、200名学生志愿者以及其他一些个体都加入了其中,他们齐心协力,一同完成了这些任务。而其间所用的建筑材料大多来自附近桥梁施工地。最终财富山被建设成为台北的一处环保型可持续发展城市实验居住区。官方认定此居住地是合法的,并将其作为公共环境艺术工作的一部分。《纽约时报》称财富山为台湾的"必游胜地",还制作了多期电视纪录片在探索频道等多个频道对其进行宣传。

### Treasure Hill

It's a place for the legitimation process of an illegal settlement of most Kuomintang veterans living on urban farming, renting shelters for migrating workers and harvesting waste from the surrounding city.

In 2002, I wrote to the Taipei City Government that they will die. This was mainly because of ecological reasons. When the modern man loses his connection to the nature, he will die.

They replied to me after one year and I was invited to Taipei by this city government to do some sort of alternative ecological urban planning, in order to avoid them from dying.

I got a lot of data from the authorities, with which I could do nothing. The human on the Taipei streets was very different than this official information, which was more like a manual of a machine. People on the streets were very humane, alive and active. They

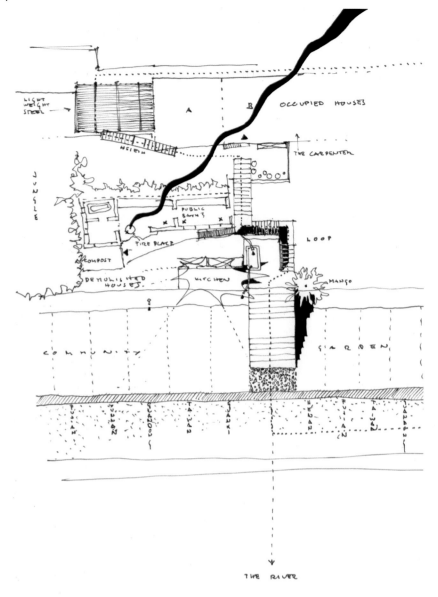

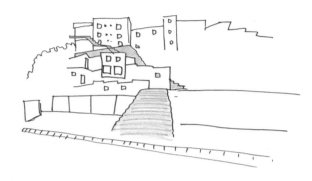

项目名称：Treasure Hill
地点：Taipei City, Taiwan, China
建筑师：Marco Casagrande
合作者：Tamkang University Department of Architecture, Taiwan University Department of Sociology, GAPP, Taipei City Government Department of Cultural Affairs
用地面积：150,000m²
竣工时间：2003

seemed to keep the industrial machine of the city alive – they were ruining it and balancing the official pollution.

I was taken to Treasure Hill, which was under a process of demolition by the same city government that had invited me to Taipei. The first three levels of the terraced houses were already bulldozed away. All the urban farms and collective gardens had been destroyed from the way of a public park to become. Treasure Hill was a high density settlement of urban farmers – old KMT veterans from the Mainland China.

What seemed to me pretty striking was the facts that this illegal settlement had many of the values, and that the official city had commissioned me to introduce for the modern Taipei: the urban farmers were filtering their grey water through vegetation, and they were cultivating passive solar houses with bioclimatic air circulation and they composted their organic waste and used a minimal amount of electricity – there was no TVs but a collective small cinema, and no cars. The urban nomads produced their own food and harvested the surrounding city from what was called waste. Treasure Hill had no problems with the flooding river, as the industrial city had built a 12 meters high reinforced concrete wall to separate the mechanical human narrative away from nature.

I changed my program, so that I focused on stopping the demolition process of Treasure Hill and actually started to rebuild the destroyed connections between the houses and restored their farms. All the work was done together with the resident and altogether 200 voluntary students and other individuals. Construction material was mostly donated from a near-by bridge construction site. Treasure Hill was changed into a laboratory of environmentally sustainable urban living in Taipei. Officially the settlement got legalized as part of a public environmental art work. According to *New York Time* the Treasure Hill is a "must see destination" in Taiwan and it has hosted various TV-documentaries including Discovery Channel. Marco Casagrande

   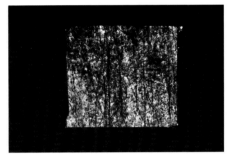

# 波将金公园

### 后工业时代的冥想公园

对于我来说，波将金公园是改革的出发点。精神上，我始终与波将金公园保持着某种联系，同时我也一直在寻求改革对我产生影响的地方。波将金公园坐落在现代人类必须明确自身与自然的关系的交叉路口。我们拥有人类在这技术世界上可持续生存所需的所有工具。而现在我们也拥有毁灭这个世界所需要的所有工具。因此建筑师、艺术家、城市规划者、环境规划者还有人类学家必须找到他们在这个关键转折点上所应承担的职责和义务。

波将金公园作为一个后工业风格的寺庙，也可以被视为一座（雅典）卫城，指引人们重新思考现代人类跟自然的关系。我将波将金公园视为一个被耕种过的废品厂，坐落在古老的稻田和直通神道教神殿的河流之间。

人们在一个非法垃圾场发现了这座公园。当时这一建筑物的建筑图纸是建筑师穿着雪地鞋在雪里按照1:1的比例走出来的，当雪融化的时候便在这里建造了这座公园。越后妻有（日本）地区可能会有3m厚的积雪（所以我们充分利用了这一点）。

波将金公园像一幅富有艺术特色的铰接式拼贴画，将城市和工业废物进行重新利用，同时它又是一个工业废墟，引起人们对后工业的反思。当一切都是人类创造的东西时，废墟便成为了自然的一部分。初走进公园，人们会看到2.5cm厚的钢墙，位于一层，但继续往前，水平面高度不断下降而钢墙的高度保持不变，这样钢墙就成为了5m高的墙壁。这个墙体系统为一组老橡树创造了适宜的空间，并且产生了一系列的室内和室外空间，以及一些小一点的寺庙和庭院，而溪流在山谷里顺流而下。在河里，你可以钓到香鱼，之后可以去波将金公园将其烤熟，吃掉，然后满意地回家。波将金，这个钢铁寺庙跟位于稻田另一边的古老神道教神殿在精神上是彼此联系在一起的。这个后工业冥想公园受到了祠官的佑护，现在120名仓俣氏村民每晚在波将金公园跳圆环舞，继续着他们有着400年历史的传统。这是一个从封建社会就有的部落仪式，用来纪念英雄事迹。所有的村民都可以到公园的这个礼堂里坐在小橡木凳上（看舞蹈表演）。

靠稻田为生的仓俣村正在慢慢消失。年轻一代人已经搬到了新泻、东京或其他城市，同时有着上百年历史的传统也正在迅速地消失；而这些传统都是建立在人与自然和谐相处，即人类是自然一部分的基础上的。波将金歌颂乡土文化传统，并且为人们提供了一处能够带给人希望的工业废墟。来自城市的参观者经常在波将金住宿并且写信告诉我说他们睡得很舒服。

## Potemkin

### Post Industrial Meditation Park

For me, Potemkin is the starting point of a revolution. I am continuing mentally attached with Potemkin and seeing where the revolution is taking to me. Potemkin stands in the crossroads where the modern man has to define his relationship with the nature. We have all the tools needed for a sustainable solution of

human existence in the technological world. Now we also have all the tools to destroy the world. Architects, artists, urban planners, environmental planners and humanists must find their position and responsibility in this turning point.

Potemkin stands as a post industrial temple, the Acropolis to rethink of the connection between the modern man and nature. I see Potemkin as a cultivated junk yard situated between the ancient rice fields and the river with a straight axis to the Shinto Temple.

The park is founded on an illegal garbage dump. The building was drawn on site in 1:1 scale on snow by walking the lines with snowshoes and then built up when the snow melted. Echigo-Tsumari region may get 3 meters of snow.

The Potemkin is an artistically articulated collage of recycled ur-

ban and industrial waste, and an industrial ruin for post-industrial meditation. Ruin is when man-made has become part of nature. As one enters the park the one-inch thick steel walls are on the ground level, but while proceeding further the ground is descending, and the walls keep levelled and thus become 5 meters high. The wall system is framing a set of old oak trees and a series of outdoor and indoor spaces, smaller temples and courtyards with the final focus on the river down in the valley. In the river, you may fish your ayu-fish, grill it and eat it up in Potemkin and go home. The steel Temple Potemkin is spiritually connected to the old Shinto Temple on the other side of the rice fields. The post industrial meditation park is blessed by the Shinto priest and the 120 Kuramata villagers are continuing now their 400-year old tradition of every night circular dance in Potemkin, a community ritual memorizing a heroic act from the feudal times. All the villagers can sit on the small oak bench in the auditorium of the park.

The rice farming village of Kuramata is dying. The younger generations have moved to Nijgata, Tokyo and other cities and the traditions of hundreds of years are about to disappear very rapidly; these traditions are based on a harmonious co-existence between the man and nature – human nature as part of nature. Potemkin celebrates local knowledge and by providing an industrial ruin it is providing hope. Urban visitors are often sleeping in Potemkin and they are writing to me that they slept well. Marco Casagrande

项目名称：Potemkin
地点：Kuramata Village by the Kamagawa River, Echigo-Tsumari, Japan
建筑师：Casagrande & Rintala
项目团队：Edmundo Colon, Chris Constant, Philippe Gelard, Leslie Cofresi,
Marty Ross, Janne Saario, Jan-Arild Sannes, George Lovett, Dean Carman,
Joakim Skajaa, Sonny Madonaldo
组织者：Echigo-Tsumari Contemporaty Art Triennial 2003, Curator Sakura Iso
维度：130m(L), 5~15m(W), 5m(H)
材料：Kawasaki steel(one inch thick), recycled concrete, recycled asphalt, recycled glass, recycled pottery, river bed stones, white gravel, oak
设计时间：2002 / 竣工时间：2003
摄影师：©Dean Carman(courtesy of the architect)

# 陈宅

该住宅建于台湾北部大屯山的一片古老的日式樱花田中,设计者把它设计成船的外形,以抵御当地暴风、洪水以及炎热的恶劣气候环境。住宅被抬离地面,这样的设计是为了方便洪水从屋底流过。在室内与室外的功能轴心区内,住宅的不同空间既连接起来,又可以灵活地移动。当遭遇频繁的台风或地震时,小小的浴室和厨房便结合在一起,来稳固整个住宅的木质结构。

这种符合生物气候学设计的建筑在炎热的气候中,可以带来大屯河凉爽的微风,并且使凉风在水库和农田之间的场地内循环。在冬天,一个火炉可以用来加热和煮茶,浴室与小型桑拿房连接在一起。

住宅既不牢固也不笨重,反而是柔弱而灵活的。住宅并没有将环境关闭在外,而是给农民提供了必要的庇护所。我们所期望的便是把这座住宅建成一座废墟。德国诗人贝托尔特·布莱希特曾说:"昨晚,我梦到一场暴风雨,它摧毁了脚手架和铁质接头,而那些木质的构件却挺过风雨,保存了下来。"

## Chen House

The house is realized on an old Japanese cherry-farm in the Datun Mountains of North-Taiwan. It is designed as a vessel to react on the demanding wind, flooding and heat conditions on the site. The house is a stick raised above the ground in order to let the flood water run under it. The different spaces are connected to a flexible movement within the axis of outdoor and indoor functions. The smaller bathroom and kitchen unit acts as a kicker stabilizing the wooden structure during the frequent typhoons and earthquakes.

场地平面图
site plan

1. 住宅01
2. Datun河
3. 池塘
4. 祖母的农场

1. house 01
2. Datun River
3. Pond
4. Grandmother's farm

The bio-climatic architecture is designed to catch the cool breeze from the Datun River during the hot days and to let in the small winds circulating on the site between the fresh water reservoir and the farmlands. A fire place is used during the winter for heating and for cooking tea. In connection with the bathroom is a small sauna.

The house is not strong or heavy – it is weak and flexible. It is also not closing the environment out, but designed to give the farmers a needed shelter. With this house we were looking forward to design a ruin.

Bertholt Brecht, a German poet said *"Last night I saw a terrible storm in a dream. It ripped off the scaffoldings and crushed the iron joints. Though, what was made of wood, stayed still."*

Marco Casagrande

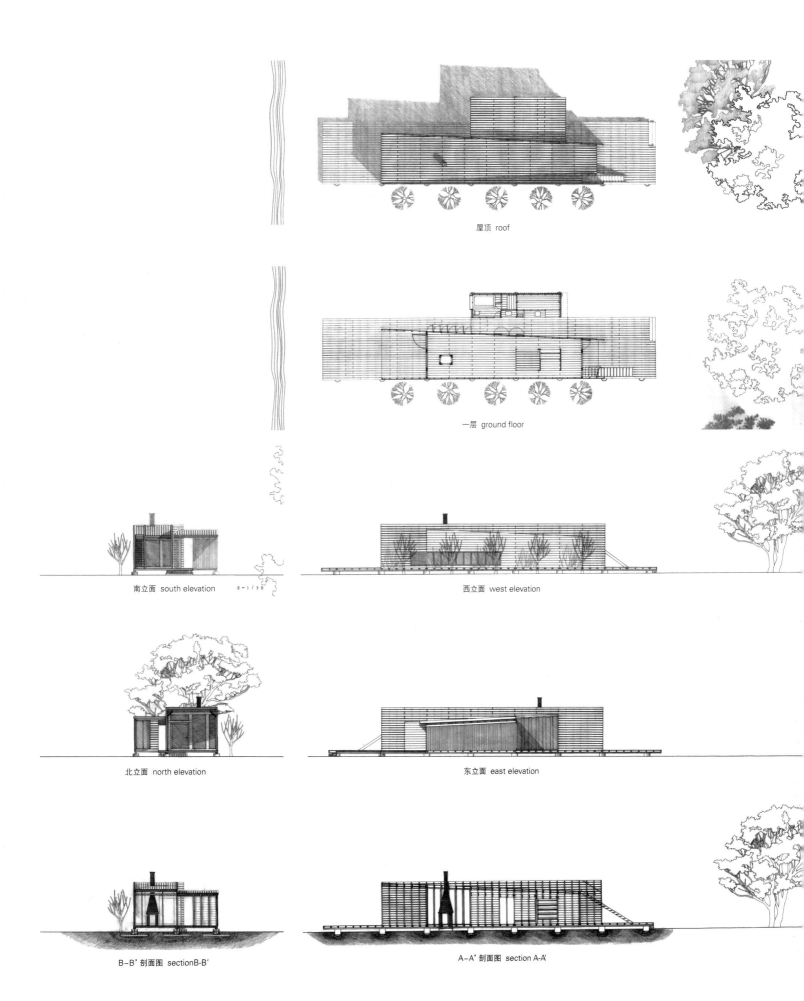

项目名称：Chen House
地点：Sanjhih, Taipei City, Taiwan, China
建筑师：Marco Casagrande
项目团队：Frank Chen, Shi-Ding Chen, Nikita Wu, Shu-Gi Bai
本地顾问：Missis Lee
用地面积：3,890m²
总建筑面积：138m²
室内空间面积：62.5m²
材料：mahogany, concrete
竣工时间：2008
摄影师：©AdDa Zei(courtesy of the architect)

# 昆虫屋

这座建筑建在一座受损建筑的废弃场地上,位于深圳市政厅和一个工人露营之间。其建筑灵感来源于昆虫。而利用竹子进行施工主要是基于地方性建筑知识,这些知识是由一些来自关西农村的农民建筑工人带来的。

在深港双年展中,这一建筑曾被用作地下乐队使用的以及举办诗歌朗诵、讨论会、卡拉OK等的多功能场所,也做过附近露营的工人的休息室。这座建筑物为他们提供了一个庇荫场所、一个舞台和一处烤火区。双年展结束后,昆虫屋将成为一个非正式的社交俱乐部,为来自中国农村的工人们服务。

在应对当地的特殊环境方面,这一建筑物显得十分柔弱,同时又柔韧灵活。由于它是从废墟中建立起来的,我们便放弃了对其进行的建筑学方面的保护以期能让自然介入其中。这样,这座岌岌可危的建筑物也就成为了人性和自然之间的中介体。这是设计者、建筑工人和地方性建筑知识之间的参与式规划的结果。这座茧状的建筑物是现代人类的一次疲软的撤退,让人们从深圳市中心爆炸式都市生活的强大力量中逃离出来。同时它又是一个避难场所,保护在非自然环境下生存的工业昆虫。

当这股改革风潮兴起时,城市再一次沸腾起来。人们必须带着已有的自由回到几千年前,这样他们才会意识到人类的生活状况并没有发生什么改变。

让一切计划的事情都实现吧。

让他们相信,并尽情嘲笑他们的热情吧。因为他们所谓的热情实际上不是情感能量,而仅仅是他们的灵魂和外部世界之间的摩擦力罢了。

最重要的是:让他们相信自己吧。让他们像孩子般无助吧,因为弱点是一个伟大的事物,而长处却一无是处。

——电影《潜行者》安德烈·塔可夫斯基

**昆虫屋有下列五个要素:**

**控制力**

意外事故往往超出人类的控制力。所以设计者必须做好充分的准备以应对意外事故的结构动态性。做好充分准备是一切艺术的关键。因此一些意外事故要求设计者们必须走出办公室(走进大自然,去进行实地考察)。由于意外事故往往超出人类的控制力,所以一些意外事故的发生使工业思想遭到了质疑,而我们通常称之为人类的错误。

**现实性**

置之死地而后生。正如一个人坐在空调屋里不可能靠近自然一样。而靠近自然的最终动机不是为了维持空调运转。

人类有一个属性:自然性。而任何其他事物也都有其各自的风格和需求法则。一只熊有毛发,是因为它是多毛类动物,并不是有人将其设计成这样。设计永远不能代替现实。真正的现实是不能被预测的;因为现实是绝对的。鱼儿不受我们意念的影响在河中逆流而上是因为鱼儿是现实的存在物。

**城市性**

工业化城市抵制自然系统进入人们的生活,所以工业化城市必须被摧毁;城市必须成为自然的一部分。在中国大量的移民正在从农村搬往工业化城市。这些人将会想出解决问题的办法,使工业化城市最终被人类的自然属性所摧毁。

**避难所**

昆虫屋是一个避难所,同时也是现代人类和自然之间的调节者。避难所是建筑学起源的种子。而昆虫屋正在从这粒种子中慢慢长大。它正在巡视四周,或许有一天它将大得超过这座城市并吞没这座城市。总有一天,所有的街道都将变得井然有序。

**戏剧性**

设计遵循着戏剧模式。社会戏剧将人类的自然属性大众化了。而昆虫屋正是一座隐藏的剧院、洞穴或寺庙,在这里建筑工人们正在日复一日地增强一股建筑改革之火。改革现在还没有完成,所以同志们,你们仍需不懈的努力。

## Bug Dome

The building is realized on a wasteland of a ruined building site in-between the Shenzhen City Hall and a workers camp. The design is inspired by insects. The bamboo construction methods are based on local knowledge from rural Guanxi brought into the city by the migrating construction workers.

The space is used during the SZHK Biennale for underground bands, poetry reading, discussions, karaoke and as a lounge for the workers from the neighboring camp. The building offers a shade, a stage and a fireplace. After the Biennale the Bug Dome will act as an un-official social club for workers from the Chinese countryside.

The building is weak and flexible to meet the site-specific conditions. It is growing from a ruin. The architectural control has been given up in order to let the nature step in. The weak architecture is a mediator between the human nature and nature. The construction is a result of participatory planning between the designers, construction workers and local knowledge.

The cocoon is a weak retreat for the modern man to escape from the strength of the exploding urbanism in the heart of Shenzhen. It is a shelter to protect the industrial insects from the elements of un-nature.

When the fire is up a society is born again. One has to take the liberty to travel a thousand years back in order to realize that the things are the same.

*Let everything that has been planned come true.*
*Let them believe. And let them have a laugh at their passions. Because what they call passion actually is not some emotional energy but just*

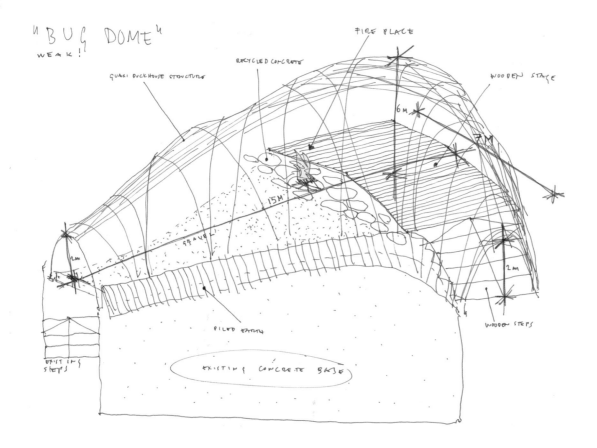

*a friction between their souls and the outside world.*
*And most important: let them believe in themselves. Let them be helpless like children, because: weakness is a great thing and strength is nothing.* – Stalker, Andrei Tarkovsky

The Bug Dome touches the following 5 elements.

### Control
Accident is greater than human control. Designer must be present in order to understand the constructive dynamics of accident. To be present is the key of all art. Accident is calling the designer to get out of the office. Accident is beyond human control, and thus insulting to the industrial mind. Usually we call this a human error.

### Reality
One has to die a bit to be reborn. One will not get closer to nature by sitting in an air-conditioned office. The ultimate reason of nature is not to keep the air-conditioning on.

Man has one identity: nature. Everything else has style and needs therapy. A bear has hair, because it is hairy – not because somebody would have designed so. Design cannot replace reality. Real reality cannot be speculated; it is total. Fish will run up to the river without our thinking on it. Fish is real.

### City
Industrial city is an anti-acupuncture needle in the life providing system of nature. Industrial city must be ruined; city must become part of nature. China has a big migration going on from the rural areas to the industrial cities. These hands can propose the solution. Industrial city will be ruined by human nature.

### Shelter
Bug Dome is in the same time a shelter and a mediator between the modern man and nature. Shelter is the seed of architecture. Bug Dome is growing from this seed. It is looking around at the surrounding city. Maybe it grows bigger and eats the city. One of these days these streets are going to get organized.

### Drama
Design follows drama. Social drama brings human nature to the street level. Bug Dome is a hidden theater, cave and a temple where the construction workers are building up a fire day after day. The revolution is not yet complete. Comrades, you should still keep working. Marco Casagrande

项目名称：Bug Dome
地点：Shenzhen, China
建筑师：Hsieh Ying-chun, Marco Casagrande, Roan Ching-yueh as WEAK!
项目经理：Nikita Wu
本地顾问：Wei Jia-kuan, Wei Jing-Ke
用地面积：300m²
总建筑面积：120m²
材料：bamboo, wood, gravel, recycled concrete
竣工时间：2009
摄影师：courtesy of the architect

去年，赫尔辛基（芬兰首都）参加了欧洲文化首都活动，而乌尼岛夏季剧院正是其中的一个项目。这座小小的临时剧院看似轻薄，坐落在那里，但却结构稳固，功能齐全，而这主要得益于其建筑材料经济且别出心裁的运用。

继萨翁林纳（芬兰湖区）非凡的乡村建筑取得成功之后，马可·卡萨格兰和萨米·林塔来在芬兰起步不久的事务所仍然继续着他们对建筑语言的探索，他们追求非永久性、经济性和非政府性的建筑理念，而这正在引起人们对特殊的环境和社会问题的关注。（例如，在萨翁林纳项目中就借鉴了空中谷仓的建筑技巧，这是对即将消失的芬兰传统的乡村生活的保护以及对郊区被入侵的反对。）作为赫尔辛基欧洲文化城市活动的一部分，这座临时的夏季剧院从该事务所的标准看似乎很传统，但是它也不乏新奇的创新和无限的活力。这一建筑坐落在乌尼岛上，属于赫尔辛基南部群岛的一部分。它是海鸟的天堂，因此一度被遗弃；而人类任何的介入都要保证将破坏降到最小值以保护其生态环境。为达到这一目的，这一建筑物的建造时间避开了鸟类的筑巢期，并且大量的构件都是在岛外进行了提前的加工，然后运到这里。

这座小剧院为一个简单的类似鼓状的亭子，外部是胶合板表皮，包裹着桦树结构。为了与周围的岩石融为一体，其外部的胶合板被涂成瓦灰色，这样建筑物看上去就像是这岛上景色的有机组成部分。圆形表演区再配上阶梯式的环绕坐椅，剧院看上去就像一个圆形马戏棚，带有节日气氛，并且还具有临时性的特质。屋顶由米白色的帆布制作而成，帆布被尽力拉伸，形成一个漏斗，可以将外部的水引向剧院内部。整座建筑都建在一个木材筏地基上，并通过一个已有的铁锚固定在场地，而这个铁锚原本是用来系船的。

剧院内部整齐划一的桦树围墙和阶梯式的坐椅，使人一进入内部就会产生一种进入乐器内部的感觉。光透过帆布屋顶柔和地射进来，照亮了整个内部空间。这座建筑物，在其短短的寿命中，将得到充分的运用，一系列的活动，例如戏剧表演、音乐会以及马戏表演都将在这里一一上演。一旦被拆除，场地将会恢复其原来的样子，而其曾经短暂的辉煌也会魔幻般消失，不会留下任何痕迹。

## Uunisaari Summer Theater

Built as part of Helsinki's program for last year's Europe's Cultural Capital, this little temporary theater sits lightly on the ground and has a robust, functional elegance that comes from economic but inventive use of materials.

Following the success of their extraordinary rural installation in Savonlina, the young Finnish practice of Marco Casagrande and Sami Rintala has continued to explore an architectural language of impermanence, economy and anarchy, that draws attention to specific environmental and social issues. (The Savonlina project,

# 乌尼岛夏季剧院

项目名称：Uunisaari Summer Theater
地点：Uunisaari, Helsinki, Finland
建筑师：Casagrande & Rintala
合作者：KUU Theatre
总建筑面积：120m² / 有效楼层面积：120m²
设计时间：2000 / 竣工时间：2000
摄影师：courtesy of the architect

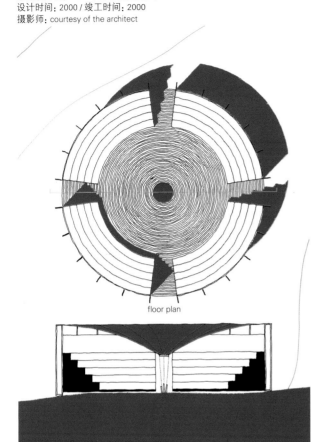

floor plan

for instance, which involved the immolation of barns on stilts, was a protest at the disappearance of Finnish traditional rural life and the encroachment of the suburbs.) This commission, for a temporary summer theater as part of Helsinki's European City of Culture programme, seems conventional by the practice's standards, but is approached with characteristically offbeat invention and vigour. The site lies on the island of Uunisaari, part of a small cluster of islands just south of Helsinki. Home to colonies of seabirds, it is normally deserted; any intervention had to minimize disruption to its ecosystem. To this end, the installation of the theater was timed so as not to coincide with the nesting season and extensive use was made of off-site prefabrication.

The little theater is housed in a simple drum-like pavilion composed of a plywood skin encasing a birch structure. The plywood is painted slate grey to merge with the surrounding rock, so the building looks as though it is an organic part of the landscape. With its tiers of seating enclosing a circular performance area, the model is clearly the circus tent, with its festive associations and quality of impermance. The roof is made of cream-coloured canvas, stretched tautly to form a funnel that drains water into the center of the space. The entire structure sits lightly on a timber raft foundation, moored to the site by existing iron anchors previously used to secure boats.

Inside, the homogeneous quality of the pale birch walls and seating creates the impression of being inside a musical instrument. Light percolates through the canvas roof, gently illuminating the space. During the course of its short lifespan, the theater was used for a range of activities, including drama, music and circus performances. Once the building was dismantled, the site was returned to its original pristine state, with no visible traces of its magical temporary transformation. Catherine Slessor

### 燃烧的激情

一个引人注目的建筑装置将人们的注意力吸引到芬兰农村的困境中来。位于芬兰的这个美妙建筑装置独具特色,设计者是马可·卡萨格兰和萨米·林塔来。它既是对芬兰传统景色和农耕活动的颂扬,又是对现今围绕每一处芬兰聚居地的、低密度郊区土地的增长的抗议。现代的农业方式标志着许多传统木建筑的时代的终结,这些木建筑曾经在乡村森林中平坦的景观草地边缘随处可见。那些小型的灰色草棚里储存着干草,身处酷寒环境的小动物都选择在这里过冬,而它们不幸的同伴和配偶都被屠杀,因为那里没有足够的饲料。今天,新型工业化农业结构和新型农业技术使那些古老的建筑看起来多余,所以它们或被摧毁,或倒塌下来。被遗弃的草棚中有三个"存活"了下来,建筑师将其解释为,"在它们必须打破与土壤相联系的地方,它们建起了自己的城堡并且正逐渐进军南方的城市。"建筑结构再一次集中起来,并且加强了内部结构。四根细长且去皮的松树枝干(由钢丝支撑)将建筑抬高10m——这种建筑正向南边的城市蔓延。

这种谦逊性被赋予了庄严性,甚至有些许的崇敬。这些建筑在走向灭亡,十月上旬,大量的干木材被捆起来丢入大火之中,且当时人们圈养的牲畜也被屠杀了。这里发生的一切采用多种方式被赋予了里程碑式的、史诗般的甚至是动人心弦的诠释,且只出现在电影和录像中。人们希望那三个英雄般的存在将永远留在这个国家的记忆及其对农业历史的意见之中。所有的评委会成员一致同意并推荐这个极有力的观点。

### Land(e)scape

#### Burning Passion

A dramatic architectural installation is designed to draw attention to the plight of the Finnish countryside. This most unusual finalist was the wonderful architectural installation in Finland by Marco Casagrande and Sami Rintala. It is (or was) both a celebration of the traditional Finnish landscape and farming practices and a protest against the endless growth of the low density suburbs which now surround every Finnish settlement. Modern agricultural methods have ensured the demise of many of the traditional wooden buildings seen on the edges of the meadow clearings of the forest all over the country's flat landscape. In these grey little barns, hay was stored, and animals chosen to live through winter were gathered in from the ferocious cold, and their less fortunate herd–mates were slaughtered since there was not enough fodder to keep them. Now that new industrialized farm structures and new agricultural techniques have made the old buildings redundant so they are destroyed or simply allowed to fall down. Three of these abandoned barns "were driven", the architects explained, "to the point where they have had to break their primeval union with the soil. They have risen on their shanks and are swaying toward the cities of the south." Their structures were put together again and reinforced internally. Then they were raised 10m high each on four slender legs of unpeeled pine trunks braced with steel wire - and they began to march towards the cities of the south.

The humbleness had suddenly been given majesty, even a degree of the sublime. They were marching to their deaths. In early October, cords of dry wood were assembled, and all was set on fire – just at the time when the beasts they housed would have been slaughtered too. The whole was in many ways a contemporary interpretation of monumental, poetic, moving, its only remaining presence on film and video. It is to be hoped that the heroic march of the three will stay on the nation's memories and its attitude to its agricultural past. All jury members agreed that the idea was extremely powerful, and that it must be commended.

Catherine Slessor

# 景观

项目名称：Land(e)Scape / 地点：Savonlinna, Finland / 建筑师：Casagrande & Rintala / 项目团队：JP Heikkinen, Aki Vepsäläinen, Heikki Leikola / 合作者：The Central Union of Agricultural Producers and Forest Owners / 面积：5,000m² / 材料：abandoned barn houses, pine trunks, timber / 竣工时间：1999 / 摄影师：courtesy of the architect

牡蛎人是位于中国台湾金门岛潮滩上的一件和环境息息相关的艺术品。在潮落时，这些人像站在离沙滩表面6m高的地方，潮涨时，则在3m高的海浪尖上行走。逐渐的，腿就会被牡蛎所覆盖。牡蛎人的中式帽子作为太阳能收集器，在夜晚的海面上发出光芒。

**Oystermen**

Oystermen is an environmental artwork at a tidal beach in Kinmen Island, Taiwan, China. The men are standing 6 meters high from the sand surface during low tide and are walking on water 3 meters high on top of waves as the ocean deepens on high tide. Gradually the legs will be covered by oysters. The Chinese hats of the Oystermen are working as solar energy collectors illuminating the seascape at night.

# 牡蛎人

项目名称：Oystermen
地点：Kinmen Island, Taiwan, China
建筑师：Marco Casagrande
材料：sandblasted stainless steel
竣工时间：2013.7
摄影师：courtesy of the architect

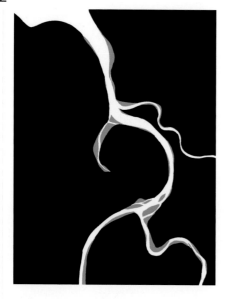
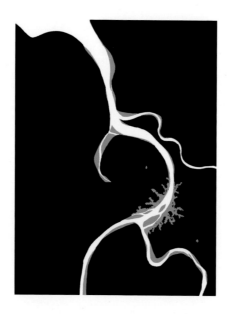
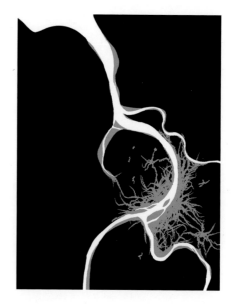

# Paracity项目

Paracity项目是一个生态有机体，以开放式形式的原则为增建基础：个性化的设计措施与周边建成的人文环境产生了广泛的交流。这种有机建构主义对话直接导致了自我管理社区结构的产生、发展及其知识建构。

这个正在发展的有机体建于一个三维木质基础结构之上，即一个带有6m×6m×6m的、由CLT层接的木棍构成的空间模块的有机网络。这种简单的结构可以由社区人员组成的团队或指定的Paracity项目建造者来进行修改和增建。

这个基本结构甚至可以在遗弃的城市地区建造，例如河水泛滥的平地、山坡、废弃的工业区、洪水河道或者贫民窟。该项目很适合洪水和海啸高发区，而CLT基础结构拥有高标准的抗震性能。

人们将他们的自建房屋、花园和农场都依附于这种基本结构，此结构为自建建筑提供了三维空间建筑网格。这种基本结构虽然提供了河流干道和流线，但是当地强有力的网络仍是居民运用自己的当地知识建立的。Paracity模型内大部分被荒野和耕地所占领。

Paracity项目的增建具有自我可持续性，且由脱离网格的环境技术方案支持，这种新技术方案包括提供了河水净化、能源生产、有机废物治理、废水净化和污泥回收利用的处理方法。这些模块化的插件可以根据Paracity的扩建来进行合适的调整，此外整个Paracity的设计不仅能够治理和循环自身的物资流，甚至还会从主体城市过滤废弃物，从而产生类似于中间的贫民窟和周边城市之间的共生方式，进而变成一个充满活力的城市寄生区。在某种意义上，Paracity是高科技的贫民窟，它可以使工业化城市朝着更加可持续的生态方向发展。

Paracity是第三代城市，是一架有机的机器，是城市的堆肥机，可以帮助人类的工业化城市转型成大自然的一部分。

### Paracity/台北

Paracity的试点项目建在台北市淡水河的一个城区耕作小岛。该岛位于中兴桥和忠孝桥之间，长1000m，宽300m。台北Paracity项目正在采用高水平的违法建筑、自我管理社区、城市农耕、社区园林、城市流浪者以及混乱不堪的构造来展现其原始的第一代城市。

台北Paracity大部分由生物能供电，这种生物能使用的有机废物，都来自周围工业城市的沉淀物，此外也由台北河流系统防洪堤内迅速产生的大量生物量供电。

驳船也应用了环境技术构件，且与Paracity城市维护码头相连接。驳船根据增长的城市有机需求可以进行改造。Paracity以通海为基础，因此没有防洪墙。虽然地上6m高的空间内没有建筑，但是整个项目屹立于支柱之上，且为社区活动、自然以及需要循坏利用院子的空间提供了整个地面层。

受集体主义意识的影响，台北Paracity将建成一个后工业化的安乐窝。据估计，这种城市模型将容纳15 000~25 000名居民。

## Paracity

Paracity is a biourban organism that is growing on the principles of Open Form: individual design-build actions generating spontaneous communicative reactions on the surrounding built human environment and this organic constructivist dialog leading into self-organized community structures, development and knowledge building.

The growing organism, the Paracity, is based on a three dimensional wooden primary structure, an organic grid with spatial modules of 6×6×6 meters constructed out of CLT cross-laminated timber sticks. This simple structure can be modified and grown by the community members working as teams or by an assigned Paracity constructor.

The primary structure can even grow on neglected urban areas, such as river flood plains, hillsides, abandoned industrial areas, storm water channels or slums. Paracity suites perfectly to flooding and tsunami risk areas and the CLT primary structure has a high standard of earthquake performance.

People will attach their individual self-made architecture, gardens and farms on the primary structure, which offers a three dimensional building grid for the DIY architecture. Primary structure offers the main arteries of water and human circulation, but the networks are grown by the inhabitants through their local knowledge. Large parts of the Paracity are occupied by wild and cultivated nature.

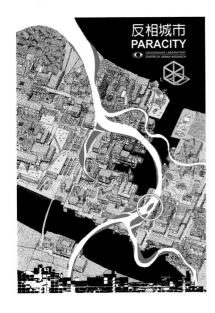
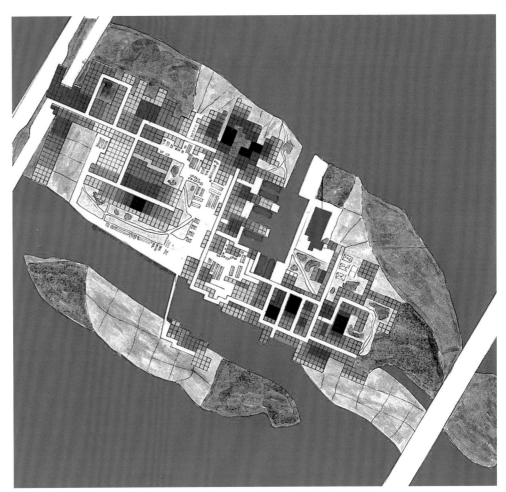
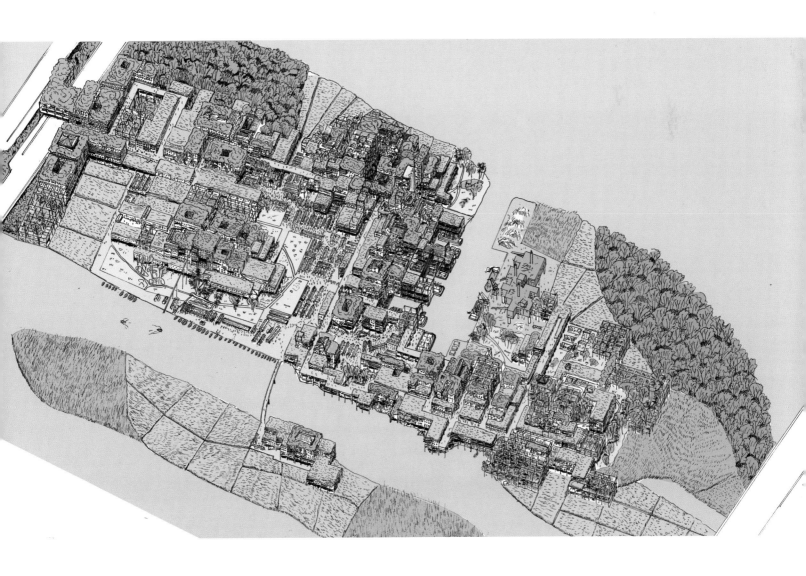

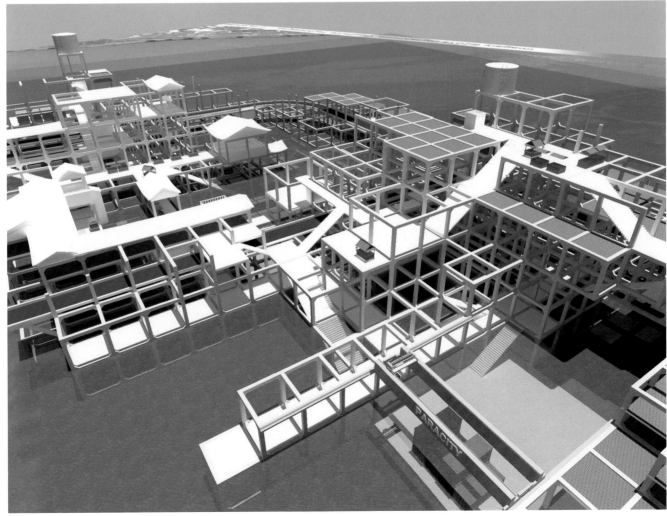

Paracity's self-sustainable growth is backed up by off-the-grid environmental technology solutions providing methods for water purification, energy production, organic waste treatment, waste water purification and sludge recycling. These modular plug-in components can be adjusted according to the growth of the Paracity and moreover, the whole Paracity is designed not only to treat and circulate its own material streams, but to start leaching waste from its host city becoming a positive urban parasite following the similar kind of symbiosis as in between slums and the surrounding city. In a sense Paracity is a high-tech slum, which can start tuning the industrial city towards more ecologically sustainable direction.

Paracity is a third generation city, an organic machine, the urban compost, which is helping the industrial city transform into being part of nature.

### Paracity / Taipei

The pilot project of the Paracity is growing on an urban farming island of Danshui River, Taipei City. The island is located between the Zhongxing and Zhonxiao bridges and is around 1000 meters long and 300 meters wide. Paracity Taipei is celebrating the original first generation Taipei urbanism with high level of illegal architecture, self-organized communities, urban farms, community gardens, urban nomads and constructive anarchy.

Paracity Taipei will be powered mostly by bioenergy that is using the organic waste, including sludge, taken from the surrounding industrial city and by farming fast growing biomass on the flood banks of the Taipei river system.

Environmental technology components are mounted on barges that are plugged into the Paracity maintenance docks. Barges can be modified according to the needs of the growing biourbanism. Paracity is based on free flooding. There are no flood walls. The first 6m level above the ground is not built, but the whole city is standing on stilts and thus providing the whole ground floor for community actions, nature and space requiring recycling yards. Paracity Taipei will construct itself through impacts of a collective conscious as a nest of post-industrial insects. Paracity is estimated to have 15,000 to 25,000 inhabitants.

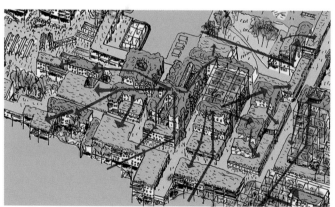

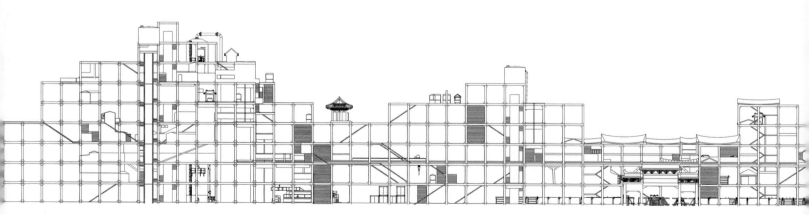

项目名称：Paracity
地点：Taipei City, Taiwan, China
建筑师：Casagrande Laboratory Center of Urban Research
项目团队：Marco Casagrande, Menno Cramer,
Katie Donaghy, Niilo Tenkanen, Nikita Wu, Joni Virkki,
Ycy Charlie, Sauli Ylinen, Dave Kan-ju Chen
用地面积：300,000m²
设计时间：2014

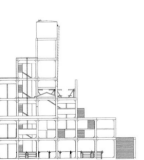

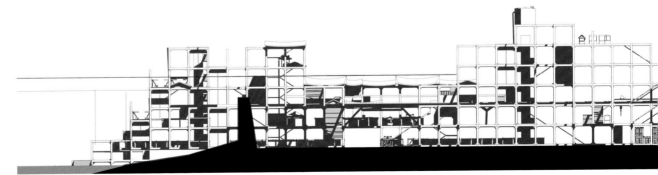

## >>20

### Renzo Piano Building Workshop

While studying at Politecnico of Milan University, Renzo Piano worked in the office of Franco Albini. After graduating in 1964, he started experimenting with light, mobile, temporary structures. Between 1965, and 1970, he went on a number of trips to discover Great Britain and the United States. In 1971, he set up the Piano & Rogers office in London together with Richard Rogers. From the early 1970s to the 1990s, he worked with the engineer Peter Rice. Renzo Piano Building Workshop was established with 150 staff in Paris, Genoa, and New York.

## >>66

### Atelier Oslo

Was established in 2006 and operated by four partners, Nils Ole Bae Brandtzaeg, Thomas Liu, Marius Mowe and Jonas Norsted. The development of each project focuses on creating architecture of high quality in which the basic elements of architecture such as structure, materiality, light and space are particularly emphasized and reinterpreted in order to solve current challenges. Its portfolio includes projects ranging from larger projects to single family houses and small installations. Their works were exhibited at the 2012 Venice Biennale and the 2013 London Festival of Architecture. And they were nominated for Mies van der Rohe Awards, European prize for Urban Public Space and Architizer Award.

## >>76

### Pezo Von Ellrichshausen Architects

Is an art and architecture studio established in Concepcion, southern Chile in 2002 by Mauricio Pezo[left] and Sofia von Ellrichshausen[right]. They was the curators of the Chilean Pavilion at the 11th Venice Biennale in 2008. Also in 2010, they were invited by Kazuyo Sejima to the official selection at the 12th Venice Biennale. They teach regularly in Chile and have been visiting professors at the University of Texas, Austin and at Cornell University. Their works have been edited in monographic issues of A+U, 2G and ARQ.

## >>106

### Häkli Architects

Seppo Häkli[left] was born in Kuusjärvi, Finland in 1951 and graduated from the TUT(Tampere University of Technology) in 1978. After working in several architecture offices including Hyvämäki-Karhunen-Parkkinen Architects, he has jointly established Häkli & Karhunen Architects in 1989 and became the owner in 1994. Was awarded honorable mention for several times at the Wood Prize Finland. Was in charge of wood studio at the Technical University of Helsinki. Matti Tervonen[right] took part in the Villa Bruun project as a building architect.

## >>124

**Casagrande Laboratory**

Marco Casagrande is a Finnish architect, environmental artist, social theorist and professor of architecture. Born in 1971 in Turku, Finland, he graduated from the Helsinki University of Technology Department of Architecture in 2001. Has been lecturing, running workshops and being in charge of design and research courses and studios in all together 55 universities in 21 countries on disciplines of environmental art, architecture, landscape architecture, urban design, sociology and on multidisciplinary courses. The universities include the Tokyo University Tadao Ando Laboratory, Aalto University Department of Environmental Art, Helsinki University of Art and Design and Bergen School of Architecture. His works have been exhibited in the Venice Biennale, London Architecture Biennial, World Architecture Festival, World Design Expo and awarded in the Architectural Review's Emerging Architecture, Mies Van Der Rohe Award, European Prize for Architecture and International Committee of Architectural Critics CICA Award.

## >>98

**Bohlin Cywinski Jackson**

Is an American architecture firm founded in 1965 by Peter Bohlin and Richard Powell. Ten principals including Ray calabro, Robert Miller and staff of 190 practice architecture, planning and interior design work including Kyle Phillips in the office. Has received the American Institute of Architects Architecture Firm Award, the most prestigious honor bestowed upon an architectural practice by the Institute. Energy efficiency and environmental sensitivity have been integral to the firm's design culture. View sustainable design as not only

### Angelos Psilopoulos

Studied architecture at the School of Architecture, Aristotle University of Thessaloniki(AUTh), then moved on to his Post-Graduate studies at the National Technical University in Athens(NTUA). Is currently pursuing his Ph.D. at the NTUA on the subject of Theory of Architecture, studying gesture as a mechanism of meaning in architecture. Has been working as a freelance architect since 1998, undertaking a variety of projects both on his own and in collaboration with various firms and architectural practices in Greece. Since 2003, he has been teaching Interior Architecture and Design in the Department of Interior Design, Decoration, and Industrial Design at the Technological Educational Institute of Athens(TEI).

### Diego Terna

Received a degree in architecture from the Politecnico di Milano and has worked for Stefano Boeri and Italo Rota. Has been working as critic and collaborating with several international magazines and webzines as editor of architecture sections. In 2012, he founded an architectural office, Quinzii Terna together with his partner Chiara Quinzii. Currently is an assistant professor of Politecnico di Milano and runs his personal blog L'architettura immaginata (diegoterna. wordpress.com).

>>88

**Dunn & Hillam Architects**

Is a Sydney-based design and architecture team operated by Lee Hillam[right] and Ashley Dunn[left]. Their projects are located all over Australia. Aim to produce honest, beautiful and practical architecture that engage with their site and location. Believe that architecture is an important tool in the creation of community, the way they interact with and create places, and the way people engage with each other. All of their work is based around the belief in the importance of a sustainable future. They place a great deal of importance on how structures are crafted and how connections are made to existing buildings and context.

>>114

**MORQ**

Was established in 2001 by Matteo Monteduro[middle], Emiliano Roia[right] and Andrea Quagliola[left]. Is a small scale architecture office dedicated to professional practice as well as university research and teaching, based in Rome, Italy and Perth , Australia since 2003. The work ranges from small residential buildings to large scale speculative projects and they are widely published internationally. Built and unbuilt projects have been awarded in international design competitions and exhibited at prestigious institutions including the Venice Biennale. MORQ is also a regular contributor at international design events and lectures at various universities.

>>32

## O'Donnell + Tuomey Architects
Sheila O'Donnell[right] and John Tuomey[left] graduated from University College Dublin(UCD) in 1976 and established their own studio in 1988. The main office is based in central Dublin and second office was opened in Cork in 2008. They have been involved in numerous projects in Ireland, the Netherlands, the UK and Hungary. They were also selected to represent Ireland in a solo exhibition at the Venice Biennale in 2004 and elected as Honorary Fellow of the AIA in 2010. They have taught at a number of European and North American schools of architecture including AA, Cambridge, Princeton, Harvard and Columbia. Both are currently teaching at UCD.

>>46

## Steven Holl Architects
Was founded in New York in 1976 and has offices in New York and Beijing. Steven Holl leads the office with partners Chris Mcvoy and Li Hu. Graduated from the University of Washington and pursued architecture studies in Rome in 1970. Joined the Architectural Association in London in 1976. Is recognized for his ability to blend space and light with great contextual sensitivity and to utilize the unique qualities of each project to create a concept-driven design. Specializes in seamlessly integrating new projects into contexts with particular cultural and historic importance. Is a tenured faculty member at Columbia University.

C3, Issue 2015.2

All Rights Reserved. Authorized translation from the Korean-English language edition published by C3 Publishing Co., Seoul.

©2015大连理工大学出版社
著作权合同登记06-2015年第19号

**版权所有·侵权必究**

**图书在版编目(CIP)数据**

林间小筑：汉英对照 / 韩国C3出版公社编；时真妹等译. —大连：大连理工大学出版社，2015.5
（C3建筑立场系列丛书）
书名原文：C3 We Live in the Wood(s)
ISBN 978-7-5611-9811-7

Ⅰ. ①林… Ⅱ. ①韩… ②时… Ⅲ. ①建筑设计—汉、英 Ⅳ. ①TU2

中国版本图书馆CIP数据核字(2015)第076295号

出版发行：大连理工大学出版社
　　　　　（地址：大连市软件园路80号　邮编：116023）
印　　刷：上海锦良印刷厂
幅面尺寸：225mm×300mm
印　张：12
出版时间：2015年5月第1版
印刷时间：2015年5月第1次印刷
出 版 人：金英伟
统　　筹：房　磊
责任编辑：许建宁
封面设计：王志峰
责任校对：高　文

书　　号：978-7-5611-9811-7
定　　价：228.00元

发　行：0411-84708842
传　真：0411-84701466
E-mail：12282980@qq.com
URL：http://www.dutp.cn